T0235736

HUMAN SUBJECTS RESEARCH

A Handbook for Institutional Review Boards

HUMAN SUBJECTS RESEARCH.

A Handbook for Institutional Review Boards

Edited by

Robert A. Greenwald
and
Mary Kay Ryan

Long Island Jewish–Hillside Medical Center
New Hyde Park, New York

and

James E. Mulvihill

University of Connecticut Health Center
Farmington, Connecticut

PLENUM PRESS · NEW YORK AND LONDON

Library of Congress Cataloging in Publication Data

Main entry under title:

Human subjects research.

Bibliography: p.
Includes index.
1. Human experimentation in medicine — Evaluation. 2. Human experimentation in medicine — United States — Evaluation. 3. Human experimentation in medicine — Law and legislation — United States. I. Greenwald, Robert A., 1943–   II. Ryan, Mary Kay. III. Mulvihill, James E., 1940–   IV. Title: Institutional review boards. [DNLM: 1. Human experimentation. 2. Professional staff committees. 3. Research — Standards. W 20.5 H918]
R853.H8H85   344.73′0419   82-5338
                                        347.304419   AACR2

ISBN-13:978-1-4684-4159-8   e-ISBN-13:978-1-4684-4157-4
DOI: 10.1007/978-1-4684-4157-4

© 1982 Plenum Press, New York
Softcover reprint of the hardcover 1st edition 1982
A Division of Plenum Publishing Corporation
233 Spring Street, New York, N.Y. 10013

"The Commission...(begins) with the premise that investigators should not have sole responsibility for determining whether research involving human subjects fulfills ethical standards. Others, who are independent of the research, must share this responsibility...."

National Commission for the Protection
of Human Subjects of Biomedical and
Behavioral Research

"The history of liberty has largely been the observance of procedural safeguards."

FELIX FRANKFURTER
Justice of the United States Supreme Court

Contributors

Joseph V. Brady, Ph.D., Professor of Behavioral Biology, Department of Psychiatry and Behavioral Sciences, The Johns Hopkins University School of Medicine, Baltimore, Maryland 21205

Jeffrey M. Cohen, Ph.D., Human Subjects Officer, Office for Research, State University of New York at Albany, Albany, New York 12222

Dale H. Cowan, M.D., J.D., Director, Division of Hematology/Oncology, St. Luke's Hospital, Cleveland, Ohio 44104; Associate Professor of Medicine, Case Western Reserve University School of Medicine, Cleveland, Ohio 44106; Lecturer, Law Medicine, Case Western Reserve University School of Law; Member, Human Research Committee, St. Luke's Hospital; Member, University Council on Human Research, Case Western Reserve University

Marien E. Evans, J.D., Attorney-at-Law; Member and Legal Advisor to the Executive Committee of the Division of Research, Lahey Clinic Medical Center, Burlington, Massachusetts 01803; Member, Board of Directors and Vice-President of Public Responsibility in Medicine and Research (PRIMR)

Myron E. Freund, M.D., Attending Urologist, North Shore University Hospital, Manhasset, New York 11030; Assistant Professor of Clinical Surgery, Cornell University Medical College, New York, New York 10021; Chairman, Research, Clinical Investigations, and Publications Committee, North Shore University Hospital

Lawrence Gold, MBA, Liaison Administrator, Long Island Jewish–Hillside Medical Center, New Hyde Park, New York 11042

Robert A. Greenwald, M.D., Chief, Division of Rheumatology, Long Island Jewish-Hillside Medical Center, New Hyde Park, New York 11042; Associate Professor of Medicine, State University of New York at Stony Brook, Stony Brook, New York 11794; Chairman, General Medical and Surgical Subcommittee, Human Subjects Review Committee, Long Island Jewish–Hillside Medical Center

Albert R. Jonsen, Ph.D., Professor of Bioethics, University of California School of Medicine, San Francisco, California 94122; Member of University of California Human Subjects Research Committee, San Francisco, California

John M. Kane, M.D., Director, Psychiatric Research, Department of Psychiatry, Long Island Jewish–Hillside Medical Center, New Hyde Park, New York 11042; Associate Professor of Psychiatry, State University of New York at Stony Brook, Stony Brook New York 11794

Bruce Kay, M.S., R. Ph., Director of Pharmacy Services, Long Island Jewish–Hillside Medical Center, New Hyde Park, New York 11042; Member, Human Subjects Review Committee, Long Island Jewish–Hillside Medical Center

Louis Lasagna, M.D., Professor, Pharmacology and Toxicology, University of Rochester School of Medicine and Dentistry, Rochester, New York 14642; Editorial Board, *IRB: A Review of Human Subjects Research*

Gwen O'Sullivan, Former executive officer, Committee on Clinical Investigations, Children's Hospital Medical Center, Boston, Massachusetts 02115

Lewis L. Robbins, M.D., Psychiatrist-in-Chief (emeritus), Department of Psychiatry, Long Island Jewish-Hillside Medical Center, New Hyde Park, New York 11042; Professor of Psychiatry, State University of New York at Stony Brook, Stony Brook, New York 11794; Chairman, Human Subjects Review Committee, Long Island Jewish–Hillside Medical Center

Martin Roginsky, M.D., Chief, Division of Endocrinology and Metabolism, Nassau County Medical Center, East Meadow, New York 11554; Associate Professor of Medicine, State University of New York at Stony Brook, Stony Brook, New York 11794; Member, Nassau County Medical Center Grants and Research Committee

Mary Kay Ryan, M.A., Research Grants Manager, Long Island Jewish-Hillside Medical Center, New Hyde Park, New York 11042; Member, Human Subjects Review Committee, Long Island Jewish-Hillside Medical Center

Barbara Stanley, Ph.D., Assistant Professor of Psychology, Department of Psychiatry, Wayne State University School of Medicine, Lafayette Clinic, Detroit, Michigan 48200

Lawrence Susskind, Ph.D., Associate Professor, Department of Urban Studies and Planning, Massachusetts Institute of Technology, Cambridge, Massachusetts 02139

Linda Vandergrift, Research Assistant, Department of Urban Studies and Planning, Massachusetts Institute of Technology, Cambridge, Massachusetts 02139

Stuart B. Wollman, M.D., Attending Anesthesiologist, Long Island Jewish-Hillside Medical Center, New Hyde Park, New York 11042; Associate Professor of Clinical Anesthesiology, State University of New York at Stony Brook, Stony Brook, New York 11794; Chairman, Pharmacology, Radiology, and Investigational New Drugs Subcommittee, Human Subjects Review Committee, Long Island Jewish–Hillside Medical Center

Foreword

For an increasing number of hospitals and universities the institutional review board (IRB) has become a way of life. Spurred into existence by public outcries about the unethical nature of certain modern scientific experiments, the IRB represents the most visible evidence of institutional commitment to ethical review of clinical research. However, this exponential growth of IRB activities has not occurred without growing pains. Like the Environmental Protection Agency, IRBs have had to develop procedures and standards without a clear consensus as to what would be optimal for science and society. Each IRB has perforce devised its own *modus operandi,* subject to general principles and guidelines laid down by others but still relatively free to stipulate the details of its functioning.

Thus one can applaud the general idea as well as the overall performance of IRBs without asserting that the millenium has arrived. The composition, philosophy, efficiency, responsibilities, and powers of IRBs remain topics suitable for debate. It is still possible (and appropriate) for IRB members to worry both about the propriety of their decisions and the personal costs of their service.

Examples of the difficult questions with which IRBs must grapple would include the following:

1. Since, at the initiation of the Tuskegee Project, patients with latent syphilis treated with the toxic heavy metals then considered "standard" therapy probably died sooner than did untreated syphilitics, was the *goal* (as opposed to the execution) of the trial—to document the natural history of the latent disease—unethical?
2. To what extent should certain special populations—prisoners, children, mental retardates, the insane—be "protected from themselves" and prohibited from participating in research? How should we reconcile the libertarian and paternalistic strands of our social web?
3. How much attention should IRBs pay to the "scientific merit" of research proposals?
4. Should IRBs be "policemen" as well as review groups?
5. How much cost is justifiable for running an IRB? Who should pay for such institutional expenses?

6. What is the optimal composition (in size and nature) of an IRB? How should one pick the so-called "public" members?
7. Should the deliberations of an IRB be open to the public?
8. Should it be necessary to assess the efficiency of consent procedures in communicating the important issues and risks to prospective subjects?
9. What is the proper role for an IRB in evolving criteria for "proper" compensation to research subjects, both for participation and damage sustained in experiments?
10. How should one provide required data to FDA monitors of clinical research while protecting the privacy of patient subjects?
11. What should an IRB do to encourage the submission for review of research that is often not now subjected to IRB scrutiny (such as new surgical or radiotherapeutic techniques)?
12. At what point—neither too soon nor too late in its evolution—should a new surgical procedure be subjected to a controlled trial?
13. Are there any circumstances under which research whose purpose would be rendered void by candid consent procedures should be allowed to proceed without such consent?
14. To what extent are present ethical codes or commission recommendations invalid as ethical guides for IRBs, investigators, and subjects?

This book provides perspective and guidance to those who have need to interact with the IRB process in a thoughtful and effective manner. However, there is no IRB equivalent of Holy Writ. Almost every chapter contains statements and issues about which persons of good will, experience, and common sense may disagree. The reason is simple: IRBs deal with matters of principle and procedures that are thorny and complex.

So much for caveats. I now hasten to urge the potential audience described above and in the preface that follows to read this book. It does not contain IRB laws carved on stone tablets, but it will help the thoughtful reader to transact his IRB business even more thoughtfully.

Louis Lasagna

Department of Pharmacology
University of Rochester
Rochester, NY 14642

Preface

Most of us need no reminder of the vast array of federal, state, local, agency, and commission rules and regulations that pervade our daily lives. This book addresses one small but crucial sphere of regulated activity: the functioning of institutional review boards (IRBs) in the protection of the rights and welfare of human subjects involved in research projects. It has been estimated that twenty percent of the health care dollar (which is itself an ever increasing share of the gross national product) is spent on compliance with regulations. It obviously behooves every professional who deals with the government to expedite such compliance. On the other hand, since most rules are generally promulgated by well-meaning groups with the general welfare in mind, a course must be navigated between streamlined compliance on one side and upholding the spirit of the law on the other. This book has been planned as a guide to the fulfillment of these dual objectives.

The history of human subjects review and of the regulations that have gradually evolved will be discussed in greater detail in Chapters 1 and 2. In brief, public awareness of the need for guidelines in dealing with human subjects in research is dated by many historians to the Nuremberg trials following World War II, where the abuses of Nazi "doctors" revealed the extent to which humans could be exploited in the name of "research." As biomedical research efforts expanded extensively over the subsequent decades, a code of ethics dealing with human experiments was formulated in 1964 in the Declaration of Helsinki. In 1966, the United States Public Health Service issued its first set of regulations dealing with human subjects research. In 1971, these rules were revised, and in 1974, a further revision was published that specified for the first time the nature of the review process that human subject research would have to undergo if federal funding was in any way involved in the project. These various regulations led to the creation of hundreds of review boards at virtually every institution involved with federally funded research. Public Law 93-348 also called for the establishment of the National Commission for the Protection of Human Subjects of Biomedical and Behavioral Research. This commission published a report in 1978 that served as the basis for revised federal regulations published in the *Federal Register* on August 14, 1979. After extensive public comment, the final regulations were published on January 26 and 27, 1981, again in the *Federal Register*. The procedures, recommendations, and suggestions in this book are based on this latest set of published rules; as of the time of editing of this book (summer 1981), substantial further changes are not anticipated in the immediate future.

IRBs arose to meet the dual need of guaranteeing protection of human subjects and simultaneously ensuring full compliance of both investigators and institutions with the federal regulations. It is probably safe to say that no two IRBs have been implemented the same way, that there has been little uniformity of procedure, that standards for "acceptable" consent forms are quite variable, and that hundreds of institutions have struggled with parallel problems of IRB review without access to general information that might expedite their deliberations. A survey of the performance of IRBs by Gray, published in *Science* (see bibliography), revealed a vast degree of heterogeneity in IRB composition, stringency of review, impact on revision of consent forms, etc. If one attends a conference on IRB function, it becomes quite clear that no two IRBs are identical and that most IRB staff and professionals have many unanswered questions about procedures, interpretation of the regulations, and numerous other issues concerning IRB operation.

This book is designed to fill the perceived need for an operational guide to IRB function. It is not meant to be a philosophical tome on the ethics and moral dilemmas of research involving human subjects. Rather, it is designed for the IRB member, the IRB staff person, and the researcher who is about to make a submission of a project proposal. Although the bulk of this book emanates from and is directed toward medical human subjects research, we have attempted to maintain a broad-based perspective that could apply equally well to research in the behavioral and social sciences.

One of the major changes wrought by the final regulations of January 26, 1981 was to narrow the scope of HHS regulatory responsibility. In the regulations proposed on August 14, 1979, and in those issued previously, HHS had taken upon itself a broad mandate to supervise all human subjects research activity at *any* institution receiving *any* HHS research funds, whether or not the specific project was so funded. HHS gave up that broad responsibility in 1981, limiting its purview to studies directly funded by the federal government. In January 18, 1980, the FDA issued final regulations on medical devices, placing increased review responsibilities on the IRB in a field where most committees had not previously trod. Although a superficial reading of the 1981 HHS regulations might lead one to believe that the workload of most IRBs would rapidly diminish, it seems safe to say that the FDA rules, the increasing complexity of state regulations, and the increasingly litigious nature of our society will keep most IRBs busier than they would care to be.

Throughout this book, we use the standard abbreviation IRB to refer to the local institutional review committee charged under the National Research Act of 1974 "to review biomedical and behavioral research involving human subjects." Although these committees exist under a variety of diverse names (human subjects committee, clinical investigations committee, etc.), it appears that IRB will be the accepted and uniform terminology of the future.

This book is divided into three sections: general principles, the review process, and special problems. The general principles of IRB functions discussed in Section I are based primarily on the experiences of IRB members and staff from the issuance of the 1974 regulations through the end of 1980. IRB regulations have been constantly evolving and we take the calculated risk that certain elements of this book may be outdated by the publication date, even though the January 26, 1981 rules have a semblance of finality about them. One of the major forces that molded these regulations was the report of the National Commission which was established by Congress to set ethical and procedural guidelines for the conduct of human subjects research. Thus it is fitting that we start, in Chapter 1, with a contribution on the evolution of regulatory influences on human subjects research; Dr. Joseph Brady, co-author of this chapter (with Albert Jonsen), is a former member of the National Commission.

The evolution of IRB procedures has also been influenced by developments in the legal sphere. Marien Evans, an attorney active in IRB work in the Boston area, traces in Chapter 2 the legal framework supporting IRB functioning. Included in her contribution is a discussion of the important but as yet untested area of IRB-member liability. In Chapter 3, Mary Kay Ryan describes the basic principles that should underlie the organization and functioning of an IRB. This discussion covers the source of IRB authority, the responsibilities and selection of members, and delegation of IRB review. Concluding Section I is a discussion by Jeffrey Cohen on the not insubstantial financial impact on an institution of maintaining a duly constituted and active IRB.

Section II is directed primarily toward IRB members and staff, although researchers preparing their submissions will obviously benefit from an understanding of the review process to which their work will be subject. Chapter 5 attempts to summarize the general principles of IRB review, notably the role of scientific merit, equitable selection of subjects, and the crux of the matter, assessment of risk/benefit ratio. Some comments on maintenance of credibility and the role of the lay reviewer are also included. Chapter 6 deals with the day-to-day operation of an IRB: maintaining a smooth flow of paperwork, staff communication with investigators, record keeping, handling changes in protocols, notation of IRB approval in publications, etc.

Chapter 7 is directed mainly toward the researcher. Although the consent of the subject is actually obtained in dialogue with the researcher, the instrument that certifies the nature of that consent is usually a signed piece of paper known as an informed consent. In preparing the consent form, the researcher must again walk a thin line between providing sufficient information about the risks of the research to assure that the subject has been fully informed and at the same time not compromising the ability to recruit subjects because of an unduly frightening consent form. This chapter reviews the required elements of a consent form, provides suggestions on the preparation of an appropriate consent, and suggests some standard wording that may be appropriate for many consents. The concepts of "risk" versus "no-risk" and oral versus written consent are also discussed.

There are two major areas of overlap between IRB function and the FDA: investigational new drugs and medical devices. The latter is of particular concern in view of the emphasis placed on IRB review by the 1980 medical device regulations. Research involving drugs and medical devices is covered in Chapters 8 and 9 by Mary Kay Ryan, with the assistance of hospital personnel who have worked on these problems. The final chapter in Section II deals with the controversial area of continuing review of research, i.e., auditing. In some institutions, formal audits are conducted in which staff members make site visits to the researcher and scrutinize the records; in other centers, researchers are merely asked to complete a form about the status of their project. Audit procedures must also be prepared to detect and deal with unauthorized and/or unapproved research.

Section III deals with a number of important problem areas where controversy and special difficulties may present themselves. In selecting topics for this section, we have attempted to emphasize areas of general interest to the research community. Thus we have deleted reference to research on pregnant women, fetuses, and special populations such as the retarded or prisoners. These topics are dealt with in great detail elsewhere, and the bottom line on most of them is that the restrictions have become so severe that much research has been precluded. We have, therefore, elected to place our emphasis on more common problems.

Two areas where many medical center IRBs often face considerable difficulty are projects involving children and protocols for cancer therapy. The former is discussed by Gwen O'Sullivan, a research staff officer at Boston Children's Hospital; the latter by Dale Cowan, a hematologist/oncologist who serves on the editorial board of the journal, *IRB: A Review of Human Subjects Research*. Two important problems usually arise during the evaluation of a cancer protocol. First, the risk/benefit equation must be evaluated in light of the fact that the alternative to study participation may be death, even though the side effects of the proposed treatment may appear ominous. Second, many protocols in this area are generated by central, interinstitutional consortia with multiple participant co-investigators. IRBs at the involved center then face the problem of reviewing protocols and consent forms that were formulated elsewhere and may not comply with local requirements. Dr. Cowan addresses the complexity of this situation.

The problem of research in surgery is addressed by Myron Freund, a urologist and chairman of a hospital IRB. Nowhere is the "boundary" problem of the borderline between standard and innovative therapy more controversial than in surgery. Many surgeons adhere to the philosophy that every operation is a unique event and that informed consent for the "research" aspect of an operation is, therefore, never required. The dilemmas posed by comparison of one operation to another, or of randomizing medical against surgical therapy, are perhaps unsolvable, but these issues are developed in Chapter 13 to guide IRBs that must address themselves to these questions.

Clinical trials of new drugs often engender special problems unique to such research. For example, a patient may enroll in a trial in the hope of seeking relief for a condition that has not responded well to conventional therapy. If the trial is of specified length, it may be the plan of the sponsor to terminate administration of the new agent even if the patient has done well during the formal testing period. This is a definite part of the risk/benefit equation that must be considered by the IRB. Similarly, the use of placebos in clinical trials is a further issue that often arouses consternation among IRB members. These issues are discussed in Chapter 14 by Martin Roginsky.

Completing Section III are two chapters relating to the behavioral sciences, one medical and one nonmedical. Two eminent psychiatrists and a psychologist have joined forces in Chapter 15 to deal with the issue of whether or not psychiatric patients are capable of giving informed consent. Finally, in Chapter 16, two social scientists discuss IRB regulation of research in their discipline.

There is an extensive literature on the use of human subjects in research, not all of which is directly related to IRB activity. An annotated bibliography of recommended papers and monographs follows the last chapter. Much of the work of the IRB involves paperwork, usually centering around the consent forms, but also involving much other documentation. In the appendix, we have provided samples of many documents that IRB staff and researchers may find useful. These include samples of consent forms in various disciplines, continuing review forms, samples of communications between staff and researchers, forms that must be filed with HHS, FDA, etc. For reference, a copy of the Declaration of Helsinki is also included.

ACKNOWLEDGMENTS

This work started as an in-house manual for researchers and IRB members at the Long Island Jewish–Hillside Medical Center (LIJ-HMC), a large and highly diversified teaching hospital in suburban New York City. LIJ-HMC has had an IRB since 1973 which has gained extensive experience in protocol review in virtually every area covered. As distribution of our local manual increased, we were encouraged to broaden its scope and convert it into the handbook that you are now reading. Many people have helped us in the tedious labor of editing a multiauthor volume. Particular thanks must go to Judith Sloan, the coordinator for human research at LIJ-HMC, Sandra Spurlock, the LIJ-HMC medical librarian, and to Sylvia Haimo, Dorothy Scharbach, Ludwina Pleickhardt, Esther Finkelstein, Ellen Dorf, and Paul Kaminsky for assistance in preparation of the manuscript. The editors hope that publication of this book will assist all involved in IRB review—committee members, staff, and investigators—in the diligent but expeditious conduct of their work.

Robert A. Greenwald *for the editors*

Contents

SECTION III: SPECIAL PROBLEM AREAS

ANNOTATED BIBLIOGRAPHY

APPENDICES

HISTORICAL DOCUMENTS

FEDERAL REGULATIONS

FEDERAL FORMS

SAMPLE CONSENT FORMS FOR SPECIFIC AREAS OF RESEARCH

BACKGROUND AND GENERAL PRINCIPLES

1

The Evolution of Regulatory Influences on Research with Human Subjects

JOSEPH V. BRADY AND ALBERT R. JONSEN

For at least the past two decades, ethical considerations involving experimentation with human subjects have become national issues of law and public policy. The federal government, sponsor of so much basic and clinical research in medicine, behavioral sciences, and other fundamental and applied fields, has promulgated guidelines and regulations of increasing explicitness and strictness (Frankel, 1972; Swazey, 1978). Despite efforts to provide guidelines for protocol review, informed consent, and risk/benefit assessment, many critical questions remain at every level of the academic and bureaucratic hierarchy. It is to these ubiquitous regulatory influences that the present volume is addressed, and it is fitting that some attention be directed to the historical antecedents which contributed to their evolution.

Historical Antecedents

The ethical issues raised by current usage of human subjects in research are certainly not new or unique. The first century physician Celsus expressly approved the vivisection of condemned criminals by his Egyptian predecessors, Herophilus and Erasistratus, with words which became a classic defense of all

---

JOSEPH V. BRADY ● Department of Psychiatry and Behavioral Sciences, The Johns Hopkins University School of Medicine, Baltimore, Maryland 21205.   ALBERT R. JONSEN ● Department of Bioethics, University of California School of Medicine, San Francisco, California 94122.

3

risky experimentation: "It is not cruel to inflict on a few criminals sufferings which may benefit multitudes of innocent people throughout all centuries" (Spencer, 1935–1938). In contrast, at the dawn of modern medicine, Claude Bernard, the master of the experimental method, espoused quite a different view: "The principle of medical and surgical morality consists in never performing on man an experiment which might be harmful to him to any extent, even though the result might be highly advantageous to science, i.e., to the health of others" (Greene, 1927). Between these two opposed ethical positions stood Sir William Osler, who, in testifying before the Royal Commission on Vivisection (1908), discussed Walter Reed's work on yellow fever. The dialogue between the Commissioners and Sir William summed up, in a remarkably precise fashion, what might be called the usual and customary ethics of research on human subjects at the turn of the century:

> COMMISSION: I understand that in the case of yellow fever the recent experiments have been on man.
> OSLER: Yes, definitely with the specific consent of these individuals who went into this camp voluntarily.
> COMMISSION: We were told by a witness yesterday that, in his opinion, to experiment upon man with possible ill result was immoral. Would that be your view?
> OSLER: It is always immoral, without a definite, specific statement from the individual himself, with a full knowledge of the circumstances. Under these circumstances, any man, I think is at liberty to submit himself to experiments.
> COMMISSION: Given voluntary consent, you think that entirely changes the question of morality or otherwise?
> OSLER: Entirely. (Cushing, 1925.)

In 1908, "human experimentation" was a relatively rare event in science. Osler's promotion of "clinical instruction" emphasized careful observation rather than deliberate therapeutic manipulation; the pathology laboratory rather than the bedside was the locus of research. It was only during the 1920s that the model of "investigator–clinician" was shaped. In the early 1930s, the methodological contributions of Sir Bradford Hill and Sir Ronald Fisher provided essential statistical tools for the design and analysis of clinical experiments. By the late 1930s, the professional clinical investigator had become established on the medical scene, and research had become an integral part of hospital practice (Reiser, 1978). Thus, with the experimental spirit abroad, the professors in position, the methods at hand, and the patients on the wards, Walter Reed's mosquito-infested hut in Havana was simulated in a variety of forms in teaching hospitals across the United States. Human beings and, most often, sick human beings were now the "animals of necessity" in theory and in fact.

Despite the unsavory sound of such phrases as "animals of necessity," these research developments aroused little indignation. Medicine was at the apogee of its scientific achievement; the gradual conquest of many lethal infectious diseases by antiseptic practice, by immunization, and by antibiotics, as well as the conquest of pain by anesthesia, had come about through research and experiment. These were triumphs which not only powerfully impressed the public, but also brought about undeniable benefit to suffering patients and to society. Research and its participants, including the human experimental subjects, basked in the glory reflected by this beneficial progress.

Background of Abuses

A more circumspect view of the historical context within which this "holy alliance" between medical practice and human experimentation had emerged, however, might reveal a less harmonious course of events. The emergence of scientific rationalism and the impact of its most prominent technological byproduct, the industrial revolution, engendered the fear of technology which found expression in a variety of 18th century literary and artistic themes. These works— *Tales of Hoffman, Rabbi of Prague, Frankenstein*—testified to an abiding concern about the effects of science and technology on the more "spiritual" aspects of the human condition. There are parallels to these expressions of alarm about "man playing God" in our contemporary biological revolution, such as recombinant DNA and *in vitro* fertilization. In addition, western individualism has challenged the paternalistic ethical norm which accepted "benefit" as a substitute for "consent." A contemporary and individualistic political philosophy, organized around the interpersonal contract, has provided a platform for exposing the weaknesses of a research ethic based upon social utility.

It was against this background that the world was shocked by the revelation in 1945 of the experiments carried out by German physicians on concentration camp prisoners; an unhappy link had been forged between the words "experiment" and "crime." Hardly an article written on the ethics of experimentation since that time fails to allude to these crimes and to comment on the impact they have had on views of the ethics of experimentation. One of these was an influential article by Henry Beecher which began with the words, "Human experimentation since WWII has created some difficult problems. . . ." After indicting 22 actual protocols as ethically deficient, Beecher concluded, "an experiment is ethical or not at its inception; it does not become ethical *post hoc*—ends do not justify means" (Beecher, 1966). In rapid succession, several of the following events in the United States were widely reported in the press and then became public issues: the experiments at the Jewish Chronic Diseases Hospital, in which cancer cells were injected subcutaneously into senile patients without their knowledge; studies on viral hepatitis at Willowbrook State Hospital, in which retarded children were deliberately infected; and the Tuskegee Syphilis Study, in which 300 black rural males were left untreated for diagnosed syphilis even after effective antibiotics were available. The influence of these events (Katz, 1972; PHS Report, 1973) in generating the public view of medical experimentation cannot be overestimated.

Then, in 1972, newspapers published reports that NIH-supported scientists were perfusing decapitated fetal heads in ketone metabolism studies. In the midst of the fierce debate which preceded the Supreme Court abortion decisions, this announcement provoked expressions of public outrage. Liberal activists, still incensed by the rights violations of the Tuskegee Syphilis Study (the victims were both poor and black!), and antiabortion conservatives coalesced around a single political pole—experimentation with human subjects. The demand for action and answers was forcefully communicated to those who, while not intimately familiar with the intricacies of clinical investigation and behavioral research, provided the funds to pay most of the bills, namely the members of Congress. In less than a decade, the "enormous dynamic of human experimentation to which not only the medical profession, but also the general public was heavily committed" (Jaffe, 1969) had been transformed into a public view of medical and behavioral research as "suspect activities that should require the approval of a governmentally constituted authority for each project" (Reiser *et al.*, 1977).

The National Commission

It was in this electrifying atmosphere that the legislation establishing the National Commission for the Protection of Human Subjects of Biomedical and Behavioral Research was born. This Commission was established by Public Law 93-348, signed by President Nixon on July 12, 1974. Although many tasks were assigned by congressional mandate, the Commission's principal work was to review the problems and practices associated with protection of the rights and welfare of human subjects involved in the various forms of biomedical and behavioral research sponsored by the federal government. Eleven Commissioners, from quite diverse backgrounds, spent four years investigating, conversing, debating, and sometimes despairing, about such topics as informed consent, risk/benefit ratios; selection of research subjects; the use of fetuses, children, prisoners, or the mentally infirm as subjects of research; and psychosurgery, immunization, and sterilization. Despite the complexity of its tasks, the Commission completed its charge by providing a coherent and reasonably consistent set of recommendations and guidelines covering the broad domain of research with human subjects. The work of the Commission was, in general, favorably received (e.g., Ingelfinger, 1977; Kennedy, 1978). The quality of that work is not difficult to judge since the Commission conducted all its business in public. It was the first national body to operate under the newly-passed Freedom of Information Act of 1974, and as a result, all of its meetings, hearings, and deliberations were conducted in open session. Every spoken and written word over the four years of its tenure was recorded, transcribed, and documented in the form of a published record. It published nine reports, each of which included recommendations for legislation and regulation together with extensive background material (see reports listed in bibliography at the end of this chapter).

As popularly conceived, the National Commission for the Protection of Human Subjects of Biomedical and Behavioral Research was presumed to have been generated by specific and spectacular abuses which focused public attention upon the research enterprise in general, but in particular upon biomedical and behavioral research. After examining the field in depth, as the legislation that established the Commission required, it became apparent that—while there was much to be done in the way of clarifying just what research was about and how it should be properly carried out—it was not, and is not, shot through with abuses. In fact, compared to sports, politics, and even certain forms of religious practice, the research enterprise, with all its human frailties, can probably be considered quite benign (Levine, 1977).

The issues raised by the "news story events" certainly deserved careful attention and review. However, a major problem arose in translating technical issues into public policy. The Commission devoted itself to the explicit tasks of formulating precise policy for the regulation of research, clarifying definitions, and critically evaluating the prevalent assumptions. Although the four years of study, deliberation, and decision were an extraordinary experiment in the melding of ethics and public policy, it seems pertinent to consider why "biomedical and behavioral research" (as the title of the National Commission clearly specified) had been singled out for this critical scrutiny to the virtual exclusion of more prevalent, although less visible, research endeavors with human subjects in engineering, marketing, and advertising, among others. The answer to this question can be sought in examination of a fundamental set of relationships which, if not unique to medicine and psychology, are displayed with prominence in these professions. The nature of these basic issues provides a focus for the key features of human subjects research ethics.

The "Boundary Problem"

An exemplary case is the so-called "boundary problem." Stated simply, it is not always crystal clear to either the patient/subject or to the doctor/ researcher, much less to the spectator/public, just where the practice of medicine or psychology as "helping profession" ends and the conduct of biomedical and/or behavioral research begins. The difficulty in establishing this boundary arises largely because the settings, personnel, and maneuvers that characterize these interacting domains are frequently the same. Moreover, the "medical experiment" with its primary objective of generating new knowledge rather than helping a given individual was not differentiated conceptually until the 19th century. Until that time it was embedded in the context of practice and no special ethical obligations had been formulated, as the ethic which governed medical practice (i.e., "no harm," "preserve life") was presumed sufficient for those experimental efforts generally viewed as attempts at patient benefit.

The blurring of the boundary between research and practice can be seen to have profound ethical significance. First, potential conflicts of interest arise because the patient/subject who places himself in the hands of a practitioner/ researcher does so primarily, if not exclusively, for personal benefit, whereas the doctor/investigator who accepts responsibility for invasion of the person may be looking toward the broader horizon of future benefits arising from the quest for new knowledge. Under such circumstances, there is the potential for a conflict of objectives, if not of methods and procedures. Second, when a person presents himself for individual treatment to a professional, this is traditionally a private matter between doctor and patient. There are almost no demands for public scrutiny. On the other hand, the hypotheses and uncertainties associated with the research quest for new knowledge—rather than direct individual benefit—clearly call for public evaluation and validation since the public is to be the beneficiary of such new knowledge and must, for the most part, bear the burdens (i.e., costs) of the research activity leading to its acquisition.

Definition of Research

The term "research," commonly used as though referring to a "thing," actually designates a class of activities directed toward the development of or contribution to generalized knowledge. This generalizable knowledge includes theories, principles, or relationships (including the accumulation of data upon which they must be based) that can be corroborated by the scientific methods of observation, experiment, and inference. Research activities may be undertaken to seek new knowledge, to restructure or reorganize existing bodies of information, to verify extant theory, or to apply existing knowledge to different situations. While the various scientific disciplines specify criteria for evaluating research performances within the scope of their respective domains, some components are common to all such investigative endeavors, including explicit objectives and formal procedures designed to attain these objectives. Both are commonly set forth in a research protocol.

To be distinguished from research activity is engagement in professional practice solely for the enhancement of an individual's well-being, with reasonable expectation of success as the standard. Indeed, the consequences of the "routine and accepted practice" of the helping professions have by long (and honored) tradition focused exclusively upon patient benefit. There are, of course, instances where research and practice may coexist (e.g., monitoring the effects and/or evaluating the effectiveness of treatment), but the aims and purposes, if not the methods and procedures, can, for the most part, be readily distinguished. Borderline areas such as innovative therapy and nonvalidated practice do exist. However, the absence of precision or validation upon which to base an expectation of success in practice does not of itself define research. Morally relevant concerns emerge on both sides of a dilemma posed by the potential for bad practice in the name of research on the one hand, and research interference with treatment or service delivery on the other. There is obvious need, in the best interests of patient/subject and doctor/investigator alike for clarification about which procedures are essential for treatment and which are introduced for research purposes (Belmont Report, 1978).

Human Subject. From these necessary but far-from-sufficient distinctions between research and practice an acceptable working definition of a "human subject" can also be derived, thus revealing the salient issues which have dominated efforts to protect this "endangered species." A "human subject" is defined as a person about whom an investigator conducting research with the objective of developing generalizable knowledge obtains the following: data through intervention of interaction with said person; and/or identifiable private information. The ethical conduct of research within the framework of these limits obviously requires a balancing of society's interests in developing generalizable knowledge on the one hand, and in protecting the rights of individual participants on the other. The identification and analysis of the elements that must be considered in this balancing of interest was the primary focus of the National Commission's deliberations and recommendations.

The Risk/Benefit Ratio

One aspect of research ethics which requires definition is the so-called "risk/benefit ratio." While it is clear that some balancing of costs and returns is necessary even in the domain of scientific investigation, the very use of the terms "risk" and "benefit" may be inappropriate at best and prejudicial at worst. One of the important advances reflected in the Commission's efforts to clarify the definition of research has been the emphasis upon "knowledge" as the product of such activity rather than "benefit," in the sense that the latter term is conventionally used in professional practice. The Commission repudiated the common but misleading phrase "therapeutic research." The use of the term "risk," carried over in large part from the practice context, creates the presumption that research should not be done because of inherent "harms," and that these "harms" can only be outweighed by attendant "benefits," again conceptualized from the perspective of the individual "patient." Considered from a somewhat broader point of view, all research aims at certain valued outcomes (e.g., increased knowledge, scientific understanding, and helpful practical applications) and involves certain costs, including in some instances possible harms (either individual or societal). In the face of these potential conflicts of obligation, the ethical task, simply said (but not so simply done!) is to ensure an equitable balance between these costs and returns.

However, the admonition of Bernard, "no harmful experiment even for great social gain!" confronts the advice of Celsus "harm to a few is justified by benefit for many!" Our culture does not unquestioningly accept the appeal of social benefit. All of the contemporary ethical codes regarding research with human subjects (i.e., Nuremberg, Helsinki, Department of Health and Human Services, American Medical Association, and American Pharmaceutical Association), while subscribing to a *bonum communum* defense of research, set it off against an "informed consent" requirement. Sir William Osler's view, "voluntary participation in risky experiment," prevails. Under such circumstances, the "consent doctrine" has become the dominant ethical issue in research with human subjects.

The Consent Doctrine

The primary justification for requiring "informed consent," as distinct from a "consent form" (Brady, 1979) resides in the right of individuals to self-determine the use of their own persons, independently of any considerations of costs and returns (Belmont Report, 1978). By implication, even cases which involve negligible or nonexistent costs require consent to research participation. This basic justification of the consent doctrine is a double-edged sword—it protects the right of an individual to participate in research which may not be cost-free or even harmless. In fact, on these grounds alone, an argument can be made for limiting (rather than expanding) the role of institutional review and governmental regulation of consent procedures.

A second justification for the consent doctrine focuses upon protection of the person by enhancing the subject's awareness of research objectives and procedures. At a minimum, this second justification preserves the right of research participants to make judgements in terms of their personal values rather than proceed on the assumption that an investigator's "advancement of knowledge" objectives are necessarily synonymous with "universal beneficence." The other side of the same coin, however, raises questions about the extent to which adequacy of information affords protection under conditions that involve apparent compromise of autonomy. The involvement of prisoners as research subjects provides an instructive example (see the report *Research Involving Prisoners* in the Bibliography). On the one hand, characterization of prison environments as "inherently coercive" suggests that essential conditions of voluntariness may be difficult to establish and maintain. On the other hand, respect for the individual prisoner's self-determination suggests a right to participate in research under conditions which provide for the availability of adequate information. In balancing the competing claims presented by such dilemmas, it is important to consider that all bad treatment in prisons is not the result of malevolence and/

or malfeasance. Some significant portion of the abuse liability is attributable to ignorance, and research activities can represent a powerful countervailing influence in such circumstances.

Similar boundary conflicts between the "autonomy" and "protection" justifications for the consent doctrine appear to be presented by those clinical research settings in which it is argued that the patient's interests are best protected by withholding harmful information which would normally be revealed in the course of obtaining consent. It is in this context involving questions of "competence" that issues surrounding "proxy consent" are frequently raised. The legal origins of the proxy concept, however, appear to reside not in the protection it affords the person whose proxy is exercised, but in the protection it provides for the proxy exerciser (e.g., "property rights" as in days not long past when wives and children were considered chattel). It remains an empirical question whether the "situational authority" frequently exercised under such circumstances is more protective than so-called "free and informed" consent.

These two qualifying terms "free" and "informed" express distinctions between the self-determination and protection justifications of the consent doctrine. Deterministic considerations aside, it is usually possible in practice to estimate the "degrees of freedom" in a choice situation by identifying alternative options. Infringements upon such "voluntarism" (a term preferable to "freedom" from a behavioral perspective) can be subtle in nature, however, and it is not morally sufficient merely to limit "force, fraud, deceit, duress, over-reaching, or other ulterior forms of constraint or coercion," as the extant codes and regulations require. True voluntariness depends upon the number of realistic options available and the extent of the individual's knowledge of these alternatives, and thereby hangs the tale of the second qualifier in the consent doctrine—its "informed" feature.

There is generally good agreement regarding the substantive nature of the information that should be made available to research participants on the basis of a "reasonable person" (rather than a "fully informed") criterion. No efforts have been spared to insure that investigators inform research subjects with respect to the purposes, procedures, attendant discomforts, alternatives, and, of course, their right to withdraw at any time without prejudice. The regulations and guidelines which have been promulgated in this regard provide for potent management of the investigator's "informing" performances by negatively consequating even the faintest suggestion of compliance failure. But what of the human research subject's "knowing" or "comprehending" behaviors? With rare exceptions, no provisions appear to have been made for this consideration in any of ethical doctrines which have emerged in the human research subject domain. What appears to be at issue is the "knowingness" of the subject and the evident difficulty in making determinations thereof based in whole or in part upon an evaluation of an investigator's informing behaviors. Even in the majority of cases, where (it is to be hoped!) the subject can be presumed to have taken an active role in the information transfer, the formal characteristics of the procedure frequently suggest that the subject is merely echoing information to which he/she has been exposed.

Preconsent Procedures

Assimilating information and understanding it are two different things. Those of us who "learned" the Lord's Prayer and "The Star Spangled Banner" by rote need hardly be reminded of how little "comprehension" need be involved. Since "echoing" is prevalent in many instructional settings, this type of verbal behavior can be expected to carry over to other less explicitly educational situations. It requires but a modest extrapolation to recognize the similarities between traditional educational settings and the operational features of consent procedures commonly employed for the "protection" of human research subjects ("Oh say can you . . . consent!").

In contrast, there are nonechoic verbal performances which reflect a speaker's comprehension. Such behaviors are influenced predominantly by functional relationships between the verbal "knowing" response and the environmental contingencies that it signifies. The credibility of a subject's "understanding" a research procedure (e.g., drawing a blood sample) is obviously enhanced by firsthand experience of the effects of that procedure (e.g., transient pain). Ideally, a consent procedure should ensure that the prospective subject does in fact have such an understanding of the procedures involved in participation, even though the extent to which that knowledge controls consenting behaviors may not be immediately obvious. (It is in this latter regard, of course, that voluntarism interacts prominently with the "informed" features of consent.) Nonetheless, it does seem worthwhile and feasible, at least under some circumstances, to assess a research subject's comprehension in this regard. Preconsent performances can offer an opportunity to evaluate such understanding of the critical functional relationships that characterize the investigative procedures.

Such a procedure has been developed over the past several years at The Johns Hopkins University School of Medicine. This preconsent procedure is designed to ensure comprehension of the research operations by prospective subjects. The experiments are concerned with the analysis of individual performances and social interactions under conditions that involve small groups of people living continuously in a residential laboratory for periods of 10 days to 2 weeks (Brady *et al.*, 1974). Men and women are invited to respond to announcements placed on local college bulletin boards and in area newspapers. All potential volunteers receive psychometric tests and are interviewed by the medical staff of the project. Then they are invited to participate in a preexperimental (and preconsent) informing procedure that involves several daily briefing sessions in the actual research setting.

During these periods, the prospective research subjects receive monetary rewards that are contingent upon the performance of certain procedures (e.g., operation of signal monitoring instruments, mathematical problem solving, environmental control devices, and exercise tasks); these procedures are components of the experiment in which consent to participate is being considered. In addition, each prospective research subject receives a manual of instructions detailing the experimental procedures and operation of environmental resources for guidance throughout the experiment. The consent form is offered for witnessed signing only after the candidate/subject has completed the entire preexperimental orientation and has provided evidence by performance of comprehension of the research procedures. Significantly, several prospective volunteers have declined to sign the consent form and participate in the research after completing the preexperimental informing session. Although this occurs infrequently, it does suggest that under conditions that emphasize positive consequential control of the participant's "knowing" performance, rather than aversive sanctions for shortcomings in the researcher's "informing" behaviors, the prospective subject's comprehension is enhanced. More importantly, this contingency management approach produces an important shift in the proportional degrees of protection afforded by the consent procedures (Levine, 1979; Lebacqz and Levine, 1977). Instead of emphasizing protection of the investigator (and his or her institution), this procedure offers more protection to the volunteer subject. Moreover, the procedure has served to reduce the risk of abortive experiments under circumstances that involve substantial investments on the part of subjects and investigators alike.

It is, of course, self-evident that the elaborate consent procedure described in this example will be neither feasible nor appropriate for most research involving human subjects. The case in point does serve, however, to highlight the distinctions between the consent form and informed consent. No amount of descriptive detail or contractual small print could possibly incorporate the warranties of comprehension that the procedure itself guarantees. At best, the consent form provides documentation of the occurrence of these informing procedures along with the more conventional statements about purposes, discomforts, alternatives, and, of course, withdrawal options. To the extent that it is executed under appropriate conditions, the consent form provides reasonable assurance of respect for personal autonomy. With regard to the protection afforded by an essential understanding of the research procedures, however, the consent form per se provides few if any guarantees. Under the circumstances, such formalities can be considered neither necessary nor sufficient conditions for compliance with the requirements of the consent doctrine.

Some Contemporary Developments

This historical review of the evolution of regulatory influences upon research with human subjects can be concluded by examination of certain broad presuppositions about research that have been prevalent among the scientific community and are shared to some extent by the public. Although these presuppositions pertain to science in general, they particularly color the kind of research which involves the use of human subjects. Recent developments such as the National Commission's deliberations have not only drawn these presuppositions into the light of critical scrutiny, but have actually challenged them quite directly. As a result, it seems likely that research with human subjects will henceforth be conducted in an atmosphere quite different from yesterday's or even today's. The differences will be apparent not merely because a new library of regulations has been created and a new cadre of instant ethicists has evolved to interpret them, but, in a much more serious sense, because the presuppositions about research and its ethics will have changed. In an implicit rather than explicit way, at least three of these presuppositions have recently been brought to the surface, exposed to a critical acid which dissolved much of their substance, and reduced to a remainder which appears in clearer view. In essence, these three presuppositions may be briefly stated as follows: first, research confers benefits upon society (and, as an ethical corollary, is justified by such benefits); second, researchers are both benevolent and trustworthy (and, as an ethical corollary, should have the right to make the principle decisions involving the use of human subjects); and third, research is not a political matter (and, as an ethical corollary, should be self-governing and self-regulating).

The first of these presuppositions, that experimentation benefits society, has been commonplace since ancient times. Indeed, the truth of this presupposition as a generalization cannot be denied, and many particular histories can be marshalled in its support. It does, however, seem timely to point out its vulnerability as a defense of research in the current climate. Public support of research has been eroded by skepticism as it has become apparent to the layman and scientist alike that translation of the products of expensive health research into benefits for patients is quite limited (Fredrickson, 1977). Not only are the benefits of research not as immediate and demonstrable as they once were, but also some dramatic breakthroughs such as the heart transplant have actually led to public disappointment. Moreover, there are increasing indications that many health problems must look more to environmental, behavioral, and life-style changes for their solution than to medical research advances of the more traditional variety. These broad statements need qualification, of course, but they do reflect the contemporary social milieu in which the "benefits of research" justification is not complacently accepted as an overriding consideration in experimentation with humans or any other kind of subject.

In the present climate, then, one cannot subscribe uncritically to the language of current regulations that requires that the risks to subjects be outweighed by the benefits to those subjects or the importance of the knowledge to be gained. The National Commission, for example, took a decidedly nonutilitarian position with regard to research in general, justifying it either by the benefit to, or voluntary consent of, the subjects themselves. As a result of its long deliberations, the Commission explicitly repudiated the position that any particular protocol could be justified principally by the "potential benefits" of the research. Rather, the necessity for scientific soundness in the investigative undertaking was emphasized, suggesting that methodological considerations relevant to the likelihood of arriving at valid conclusions were ethically more weighty than speculation about benefit. This, of course, is in good accord with the position that the product of research should be considered in terms of generalizable knowledge rather than the more remote and problematic "health benefit."

Expressions of skepticism concerning the benefits justification of research should not, of course, be dogmatic. Certainly, good arguments of this sort can and will be made, but they must be made in a discriminating manner. There is obvious need for more refined categories of discourse in which the argument spells out what the benefits are, i.e., how likely they are to come about, who will be the beneficiaries, precisely why the intended results should be called "benefits," and from whose perspective. In parallel, there must be precise categories of harm: What sorts of harm might result? Who might suffer it? How likely is it? How serious is it? Is the harm physical, psychological, social, or economic? What are the social and economic costs of the benefit and of foregoing it? These considerations have long been part of the discourse about research ethics, but they have often been vague in concept and imprecise in use. It seems well worth reiterating the peculiarities which characterize the very language of "benefits and risks" in which "benefit" is a word expressing actuality with connotations of certitude, while "risk" is a questioning word expressing only possibility. In research both benefit and harm are future and uncertain, and the call for more specific articulation of both can be heard in the Commission's deliberations and elsewhere. All too often, acknowledging the moral dimensions of research leads to the familiar lapse into vague discourse about "values" and the ultimate conclusion that such unresolvable moral issues in research must, regrettably, be set aside (Bok, 1978).

The second presupposition regarding the benevolent and trustworthy exercise of authority by researchers over their work has been eloquently and articulately defended on repeated occasions both in the past (Cushing, 1925) and the present (McDermott, 1967). Again, a cautious note of skepticism seems in order, though certainly not as an invidious indictment of all researchers as either sadistic or selfish. While benevolence and trust must continue to be highly regarded interbehavioral virtues, the science enterprise has become a vast and complex institution with resources that exert powerful control over the activities of researchers. Scientific investigation is no longer a lonely, unacknowledged sacrifice as it was for Ignace Semmelweis. Contingent consequences of research performance involve financial support, academic advancement, professional reputation, and, perhaps at the end of the rainbow, The Prize! Clearly, regardless of personal probity, there is a question of conflict of interest.

The National Commission recognized these problems and disclaimed, though not explicitly, the beneficence of the investigator as an ethical justification of research. This was done by employing the devices which common law has long used to deal with conflict of interest: full disclosure and the intervention of third parties. Informed consent and institutional review are, of course, less than perfect remedies, but they are the pillars of ethical research in our society and they must be taken seriously (Gray *et al.*, 1978). We cannot concede, as many of our research colleagues have asserted, that consent can never be informed and voluntary or that reviewers can never adequately judge the protocols of specialists. Consent and review must be taken seriously because they are the only means of acknowledging conflicts of interest, rendering them public, and thus impotent as threats to research subjects.

The final presupposition that research is not a political matter and thus should be immune from political pressure, influence, and control would hardly seem to require extensive rebuttal in the face of current realities. It was the mutual interests of researchers (in the political process and its largesse) and politicians (in both the public benefits and public relations value of research) which generated the enormous investments in our National Institutes of Health. Still, the grant process was originally designed with great care to shield research from the inconstant interests emanating from the legislature, and many researchers seemed to live for a time with the innocent and uninformed belief that science was as separate from politics as church from state. It is now painfully evident, however, that the issues that made at least the ethics of research a public matter were thoroughly political. Tuskegee, for example, was racial, and fetal experimentation was related to abortion, two issues which generated immense political energies. Other current questions involving research ethics make contact with potent political energy sources focused upon women's rights and the environmental movement. The problems of research have been shown to reside not only in the laboratory and the clinic, but in the Congress as well. Who among us has not dreamed (or had nightmares!) about the prospect of receiving a Golden Fleece Award?

In reflecting upon the merits of such contemporary reevaluations of traditional views as they relate to human subjects research, two points seem worthy of emphasis. First, in an era of ethical pluralism acknowledged as both a fact and a value, individuals representing quite different points of view have been able to reach broad consensus about many critical problems in research ethics. Although frequently disagreeing on "principles," sustained consideration and deliberation can produce substantial agreement about "conclusions." Responsible groups of individuals faced with a specific task can produce not the expected "pluralism," but in fact, genuine practical agreement upon a reasonably coherent and consistent set of recommendations and guidelines. Second, in the domain of ethical judgment, where it has become axiomatic to regard empirical findings as necessary but not sufficient, no effort must be spared to acquire extensive factual data about mandated problems. Systematic inquiry—biomedical, behavioral, sociological, legal, and philosophical—can be both productive and demonstrably useful. And though all of this factual information may be indispensible in consideration of "value" questions, the nature of such "research in ethics," especially for public policy purposes, remains unclear. Despite this lack of clarity, such research information must be sought, received, and usefully considered. As public policy questions of increasing import continue to engender ethical debate, it would seem worthwhile for the community of scholars and the body politic to explore further such methodological approaches to "experimental ethics."

ACKNOWLEDGMENTS

Research referenced in this paper was supported by Grants DA-00018 (NIDA) and NGR-21-001-111 (NASA).

References

Beecher, H. K., 1966, Ethics and clinical research, *N. Engl. J. Med.* **27**:1354–1360.

Belmont Report, 1978, *Ethical Principles and Guidelines for the Protection of Human Subjects of Research,* DHEW Publication No. (OS) 78-0012, Vol. 1, Appendix, pp. 2–4, U.S. Government Printing Office, Washington, D.C.

Bok, S., 1978, Freedom and risk, *Daedulus* **107**(2):115–127.

Brady, J. V., 1979, A consent form does not informed consent make, *IRB: Review of Human Subjects Research* **1**(7):6–7.

Brady, J. V., Bigelow, G., Emurian, H., and Williams, D. M., 1974, Design of a programmed environment for the experimental analysis of social behavior, in: *Man–Environment Interactions: Evaluations and Applications,* Vol. 7, *Social Ecology* (D. H. Carson, ed.), pp. 187–208, Environmental Design Research, Milwaukee.

Brady, J. V., and Emurian, H. H., 1978, Behavior analysis of motivational and emotional interactions in a programmed environment, in: *Nebraska Symposium on Motivation* (R. Dientsbier and H. Howe, eds.), pp. 81–122.

Cushing, H., 1925, *The Life of Sir William Osler,* Vol II, p. 109, Clarendon, Oxford, England.

Frankel, M., 1972, The Public Service Guidelines Governing Research Involving Human Subjects: An Analysis of the Policy Making Process, George Washington University, Washington, D.C.

Frederickson, D. S., 1977, Health and search for new knowledge, *Daedulus* **106**(1):159–170.

Gray, B. H., Cooke, R. A. and Tannenbaum, A. S., 1978, Research involving human subjects, *Science* **201**:1094–1101.

Greene, H. C. (translator), 1927, *C. Bernard, An Introduction to the Study of Experimental Medicine* Henry Schumann.

Ingelfinger, F., 1977, Ethics of human experimentation as defined by national commission, *N. Engl. J. Med.,* **297**:44–46.

Institutional Review Boards, 1978, DHEW No. (OS) 78-0008, U.S. Government Printing Office, Washington, D.C.

Jaffe, L. J., 1969, Law as a system of control, in: *Experimentation with Human Subjects* (P. A. Freund, ed.), pp. 197–206, Braziller, New York.

Katz, J., 1972, *Experimentation with Human Beings,* pp. 10–65, pp. 1007–1010, Russell Sage Foundation, New York.

Kennedy, E., 1978, The Congressional Record S9690, June 29.

Lebacqz, K., and Levine, R. J., 1977, Respect for persons and informed consent to participate in research, *Clin. Res.* **25**:101–107.

Levine, R. J., 1977, Non-developmental research on human subjects: The impact of the recommendations of the National Commission for the protection of human subjects of biomedical and behavioral research, *Fed. Proc.* **36**:2359–2363.

Levine, R. J., 1979, Changing federal regulations of IRBs: The Commission's recommendations and the FDA proposals, *IRB: A Review of Human Subject Research,* **1**:1–3.

McDermott, W., 1967, Opening comments to Part II of Colloquium on Ethical Dilemmas from Medical Advances, *Ann. Int. Med.* **67**: 39–42.

Reiser, S. J., 1978, Human-experimentation and convergence of medical research and patient-care, *Ann. Am. Acad. Polit. Soc. Sci.* **437**:8–18.

Reiser, S. J., Dyke, A. and Curran, W. (eds.), 1977, *Ethics in Medicine,* p. 257–277, MIT Press, Cambridge.

Spencer, W. G., 1935–1938 (translator) *A. G. Celsus, De Medicina,* W. Heineman, London and Harvard University Press, Cambridge.

Swazey, J., 1978, Protecting the "animal of necessity": Limits to inquiry in clinical investigation, *Daedulus* **107**(2):129–145.

Bibliography

Disclosure of Research Information, 1977, DHEW Publication No. (OS) 77-0003, U.S. Government Printing Office, Washington, D.C. Ethical Guidelines for Delivery of Health Services, 1978, DHEW Publication No. (OS) 78-0010, U.S. Government Printing Office, Washington, D.C.

Psychosurgery, 1977, DHEW Publication No. (OS) 77-0001, U.S. Government Printing Office, Washington, D.C.

Public Health Services Report, 1973, Tuskegee Syphilis Study *Ad Hoc* Advisory Board, U. S. Government Printing Office, Washington, D.C.

Research Involving Children, 1977, DHEW Publication No. (OS) 77-0004, U.S. Government Printing Office, Washington, D.C.

Research Involving Mentally Infirm, 1978, DHEW Publication No. (OS) 78-0006, U.S. Government Printing Office, Washington, D.C.

Research Involving Prisoners, 1976, DHEW Publication No. (OS) 76-131, U.S. Government Printing Office, Washington, D.C.

Research on the Fetus, 1976, DHEW Publication No. (OS) 76-127, U.S. Government Printing Office, Washington, D.C.

The Legal Background of the Institutional Review Board

MARIEN E. EVANS

The past decade has seen the development of a new area of law dealing with biomedical research. While impetus for such a development was provided by federal and state efforts to regulate human research activities, judicial concern with human experimentation has been apparent since the eighteenth century. This chapter will review some of the legal background that underlies the current IRB regulations.

Historical Overview

In a 1767 medical malpractice action, *Slater vs. Baker vs. Stapleton,*[1] the King's Bench addressed the issue of experimentation. In that case, the defendants attempted to straighten the patient's leg after a fracture by using a device that they had developed and that was not customarily used by physicians of the time. In holding the defendants liable for the harm suffered by the patient, the court stated that:

> It appears from the evidence of the surgeon that it was improper to detach the callous (which had formed during the healing process) without consent: this is the use and law of surgeons; then it was ignorance and unskillfulness in that very particular, to do contrary to the rule of the profession, what no surgeon ought to have done; and indeed it is reasonable that a patient should be told what is about to be done to him, that he may take courage and put himself in such a situation as to enable him to undergo the operation. . . . This was the first experiment made with this instrument; and if it was, it was a rash action: and he who acts rashly acts ignorantly. . . .[2]

MARIEN E. EVANS • Lahey Clinic Medical Center, Burlington, Massachusetts 01803.

Thus, Slater defined experimentation as being treatment that is not standardly administered, and he established the requirement of obtaining informed consent to such experimentation.

In an 1871 New York case *(Carpenter v. Blake)*,[3] the Supreme Court of New York enunciated the basic dilemma that confronts biomedical researchers when it sustained a jury award to a patient whose dislocated elbow had been treated unsuccessfully in an unorthodox manner. The court, squarely facing the risk of adhering to orthodox treatment, stated:

> It must be conceded that if a surgeon is bound, at the peril of being liable for malpractice, to follow the treatment which writers and practitioners have prescribed, the patient may lose the benefits of recent improvements in the treatment of diseases or discoveries in science, by which new remedies have been brought into use; but this danger is more apparent than real. One standard, by which to determine the propriety of treatment, must be adopted; otherwise experience [sic] will take the place of skill, and the reckless experimentalist the place of the educated, experienced practitioner.

Carpenter held that the physician who departs from established medical procedures, albeit with the best of intentions, will be held liable for all harm resulting to the patient.[4]

The *Slater* and *Carpenter* decisions exemplify not only judicial thinking but also the approach to biomedical research as it continued until the twentieth century. Before World War II, biomedical research was rarely conducted in other than an ad hoc fashion. Organized efforts were rarely undertaken and those efforts were characterized by a lack of concern for the well-being or protection of the rights of the subjects of the investigation.

One example of such research is the Tuskegee Syphilis Study initiated by the United States Public Health Service in the 1930s. That study, which became infamous in the 1970s, was designed to discover the long-term effects of untreated syphilis. Two groups of subjects were involved, one suffering from the disease and the other deemed syphilis-free. No treatment of the disease was provided to either group—not even after the discovery of antibiotic therapy. Although other medical services were provided to the participants, efforts were made to prevent them from obtaining treatment for syphilis from other sources.

According to an ad hoc advisory panel appointed by the Assistant Secretary for Health in 1973 to study the project, it appeared that informed consent had not been obtained from the participants and that standardized evaluation measures had not been utilized. Subsequent litigation on behalf of the participants has reportedly resulted in settlements ranging from a high of $37,500 for survivors who had syphilis to $5000 for the estates of deceased participants who had not contracted the disease (Hershey and Miller, 1976).

The first attempt of the law to deal with the problems of modern biomedical research was the prosecution of Karl Brandt and others by the Nuremberg Military Tribunal. As there were no standards by which the conduct of the accused could be judged, a set of principles was developed for the tribunal. The Nuremberg Code, as it came to be known, embodied moral, legal, and ethical standards for judging the human experimentation conducted by the Nazis. The tenets of the Nuremberg Code have been refined somewhat and incorporated into professional codes and federal and state laws and regulations. The ten principles are the following:

1. The voluntary consent of the human subject is absolutely essential.
2. The experiment should be such as to yield fruitful results for the good of society, unprocurable by other means or methods of study and not random and unnecessary in nature.
3. The experiment should be so designed and based on the results of animal experimentation and a knowledge of the natural history of the disease or other problem under study that the anticipated results will justify the performance of the experiment.
4. The experiment should be so conducted as to avoid all unnecessary physical and mental suffering and injury.
5. No experiment should be conducted where there is an *a priori* reason to believe that death or disabling injury will occur; except, perhaps, in those experiments where the experimental physicians also serve as subjects.
6. The degree of risk to be taken should never exceed that determined by the humanitarian importance of the problem to be solved by the experiment.
7. Proper preparations should be made and adequate facilities provided to protect the experimental subject against even remote possibilities of injury.
8. The experiment should be conducted only by scientifically qualified persons. The highest degree of skill and care should be required through all stages of the experiment of those who conduct or engage in the experiment.
9. During the course of the experiment the human subject should be at liberty to bring the experiment to an end if he has reached the physical or mental state where continuation of the experiment seems to him to be impossible.
10. During the course of the experiment the scientist in charge must be prepared to terminate the experiment at any stage, if he has probable cause to believe, in the exercise of the good faith, superior skill, and careful judgment required of him that a continuation of the experiment is likely to result in injury, disability or death to the experimental subject.[5]

Thus the Nuremberg Code provided a far more sophisticated set of principles than those of the early decisions dealing with individual cases and activities categorized as experimentation in malpractice litigation. The Code also provided modern medical researchers with standards by which their professional practices could be measured.

Regulation of biomedical research by the federal govenment followed increased federal support of such research. As early as 1953, the National Institutes of Health required that research involving humans at its Clinical Center in Bethesda receive approval by a review committee responsible for the protection of subjects prior to supporting the research. In 1966, the Surgeon General, through the power granted to the Public Health Service to regulate the conduct of research for which it provides funds, extended the requirement of prior review to all "extramural" research supported by that agency. Review of such research was conducted by committees of "institutional associates" as part of the peer review process (Gray, 1977). In 1971 the Department of Health, Education, and Welfare (DHEW) published the *Institutional Guide to DHEW Policy on the Protection of Human Subjects,* which required institutional committee review of DHEW-funded research involving humans as subjects.[6] DHEW regulations were proposed in 1974 and promulgated in final form in 1975; they remained effective through the end of 1980. Final regulations were announced in January, 1981, partly as a result of recommendations from the National Commission for the Protection of Human Subjects of Biomedical and Behavioral Research.[7]

The Food and Drug Administration, responding to the congressional mandate of the 1962 Food and Drug Amendments, [8] adopted the policy of requiring that informed consent be obtained from individuals who participate in new drug investigations. Subsequently, in 1971, the FDA promulgated regulations requiring that institutional committees review clinical studies of new drugs in which human beings serve as subjects.

Even though the Public Health Services and the Food and Drug Administration had been concerned with the protection of human subjects for several years, disclosure of a number of questionably conducted research projects aroused public attention and accelerated regulatory efforts. Among those projects were the Tuskegee Syphilis Study discussed above, the Jewish Chronic Disease Hospital case[9] and the Midgeville State Hospital investigation.

In the mid 1960s, physicians at the Jewish Chronic Disease Hospital, at the request of a physician/investigator of the prestigious Sloan–Kettering Institute for Cancer Research, injected into 22 chronically ill patients cells from human cancer tissues. A member of the hospital's board of directors sued the hospital to gain access to patients' records in order to determine the extent to which patients had been involved in the study. Subsequently, during the proceedings to revoke the licenses of the two staff physicians primarily involved, it was revealed that effective consent had not been obtained from the patient subjects and that the protocol had not been submitted to the hospital's research committee for review. The proceedings revealed that Sloan–Kettering conducted an institutional review of the project, but it was also discovered that the physicians having primary care responsibility for the patients had not been approached for their approval of the study. The licenses of the two physicians were suspended by the Board of Regents; however, the Board granted a stay of the suspension and placed the physicians on probation for a period of one year.

In 1969, it was revealed that physician participants in an investigational new drug program at Georgia's Midgeville State Hospital were conducting clinical drug investigations without obtaining the consent of the clinicians responsible for the patient subjects, from the patients themselves, or from anyone responsible for them. Additionally, the program had not been formally reviewed by the institution. A committee of the Georgia Medical Society, which had investigated staff practices at the hospital, recommended that a committee consisting of five staff physicians plus the hospital superintendent review each new drug program to assure the safety of the patients.

Additional pressure on the federal government to develop a regulatory system designed to protect human subjects of biomedical research came from judicial attention to individual rights. The decisions in the *Kaimowitz v. Department of Mental Health*[10] and *Wyatt v. Stickney*[11] cases are noteworthy in this regard.

In 1973 Gabe Kaimowitz of the Michigan Legal Services intervened on behalf of a mental patient who had agreed to participate in a study designed to evaluate the relative efficacy of psychosurgery and hormonal treatment in controlling aggression. The patient and his parents had signed consent forms agreeing to the performance of an amygdalotomy—a form of psychosurgery. Although the project had been approved by scientific and human rights committees, the court concluded that because of the experimental nature of the procedure, the high risk involved and the uncertainty of the results, the patient's consent to surgery was invalid. The doubt of the *Kaimowitz* court concerning the validity of consent to experimental psychosurgery was reflected in the 1978 regulations proposed by the Department of Health, Education, and Welfare.[12]

Wyatt v. Stickney[13] was one of the early cases that recognized the rights of the institutionalized mentally infirm not to be subjected to experimentation without their consent. As in the Willowbrook case, nontherapeutic experimentation had been conducted on institutionalized individuals. That research, seeking a vaccine for infectious hepatitis, involved infecting new arrivals with hepatitis. This project was one of the factors precipitating a consent decree that absolutely forbade medical experimentation.[14] Concern over the events that gave rise to such cases as *Wyatt* and Willowbrook resulted in the 1978 regulations governing research involving persons institutionalized as mentally disabled. These regulations imposed additional duties upon institutional review board members in order to protect the subject, and they reflect the findings and recommendations of the Georgia Medical Society in the Midgeville State Hospital investigation, as well as the requirements of the consent decrees entered into the Willowbrook and *Wyatt* cases.

All of the federal guidelines and regulations promulgated since 1971 have mandated that an institutionally sponsored and locally based committee accept responsibility for protecting the rights and safety of human subjects of biomedical research. Each revision of and addition to the regulations has increased the amount and kind of responsibility thrust upon IRB members. Initially, the task of IRB members was threefold. First, the board was required to assure that the rights and welfare of the subjects were adequately protected. The institutional boards were directed to "carefully examine applications, protocols, or descriptions of work to arrive at an independent determination of possible risks,"[16] and to assure themselves that precautions would be taken to deal with emergencies that might develop in the course of the study.

The second task was to assure that the risks to the individual were outweighed by the potential benefits to him, or by the importance of the knowledge to be gained. Among the factors to be considered in making a determination of an appropriate risk/benefit ratio was the possibility that subjects might be motivated to accept risks for unsuitable and inappropriate reasons.

The third duty of the IRB was to assure that the informed consent was obtained by adequate and appropriate methods. The basic elements of informed consent are the following:

1. A fair explanation of the procedures to be followed, including an identification of the experimental elements.
2. A description of the attendant discomforts and risk involved.
3. A description of the benefits to be expected.
4. A disclosure of appropriate alternative procedures that would be disadvantageous to the subject.
5. An offer to answer any inquiries concerning the procedures.
6. An instruction that the subject was free to withdraw his consent and to discontinue participation in the project or activity at any time.

Additionally, the agreement was not to include any exculpatory language by which the subject appeared to waive any legal rights or to release the institution from liability for negligence. The latter requirement originated in the *Tunkl v. California*[17] determination that the use of such clauses in consent forms is contrary to public policy.

Since 1971, both the FDA and the HHS (formerly HEW) have added to the elements of informed consent. Additional procedures have been set forth to be followed when certain populations of subjects are to participate in the research, for example, individuals confined to prisons,[18] pregnant women and fetuses,[19] children,[20] and the mentally infirm.[21] Upon the recommendation of the National Commission for the Protection of Human Subjects of Biomedical and Behavioral Research[22] and the President's Commission for the Study of Ethical Problems in Medicine and Biomedical and Behavioral Research, the HHS has finalized regulations amending its basic policy for the protection of human subjects.[23] HHS's new regulations, except for certain aspects that are unique to the FDA, are applicable to all research funded or controlled by HHS.

These new regulations impose greater responsibilities upon the IRB members. Among these is the task of ascertaining the acceptability of research proposals in terms of institutional commitments and regulations, applicable law and regulations,[24] as well as standards of professional conduct and practice.[25] Additionally the IRB must examine and judge the appropriateness of the proposed research design and the merit of the study in question.[26] Moreover, the new regulations invest the IRB with the explicit authority to suspend or terminate approval of research projects that are not being conducted in accordance with the IRB's requirements or that have been associated with unexpected harm to subjects, [27] as well as the responsibility for reporting investigator noncompliance to HHS's Office for the Protection from Research Risks.[28]

Liability of IRB Members

Imposition of the additional responsibilities upon IRB members carries a concomitant increase in their potential legal liability. Such liability emanates from failure of IRB members to exercise reasonable care when dealing with the investigators, the subjects, the institution, and the public.

IRBs located in public institutions, or those in private institutions whose internal rules and regulations provide for such procedures, must afford procedural due process rights to investigators whose research protocols are disapproved or modified. Such due process rights, which have been developed by the Supreme Court in recent years,[29] arise from the investigators' proprietary interests in their research protocols. Procedural due process demands that the IRB act reasonably in applying standards for the protection of subjects and only impose conditions that are reasonably related to subject protection or legitimate institutional requirements. The minimum due process to be afforded to investigators is "some kind of notice and some kind of hearing."[30] The investigator should be provided notice that adverse action is being contemplated and on what basis, and should be provided the chance to respond to the IRB's concerns before action is taken. In the case of original or continuing protocol review, the IRB would not be required to provide the full panoply of due process rights—the right to be represented by counsel and to present and cross-examine witnesses. However, more extensive procedural safeguards would be appropriate where the institution intends to impose sanctions for unethical or unprofessional conduct or for noncompliance with IRB requirements. Failure to provide an investigator with the appropriate procedural rights could render the IRB and its members subject to legal action and liable to the investigator for any damage suffered by the investigator because of their actions.

IRB members may find themselves personally liable to subjects and investigators for negligence in carrying out their reviews or for failing to monitor approved studies properly. To date, only one IRB has reportedly been sued. In the case of *Nielsen v. Regents of the University of California*[31] an IRB member sued other members of the IRB to prevent them from "approving, aiding or abetting a research project involving children." The Chairman of the University of Maryland's IRB has been sued for approving research involving prisoners who allegedly were not provided sufficient information upon which to base valid consent.[32] The principle that one who assumes the task of protecting others must act responsibly could result in the imposition of legal liability upon IRB members who fail to use reasonable care in reviewing proposals. If an IRB approves research activities that result in injury to the subjects, and if a reasonable person in possession of the same information would not have approved the activities or would have placed restrictions on the conduct of the research that would have prevented the injury, the IRB might be found liable for the injury to the subject.

In addition to providing investigators with some form of procedural due process, IRBs must act responsibly in reviewing protocols. IRB approval or nonapproval of a protocol should be based on factors germane to the protocol at hand and on the basis of the investigator's proven ability. If extraneous factors or information become the basis of its disapproval, the IRB may be found to have acted arbitrarily, capriciously, or maliciously, and hence be liable to the investigator for damages from interference with business relations. Similarly, reporting inaccurate information regarding an investigator's qualifications or his/her conduct of a study might result in a defamation suit by the damaged investigator. Finally, the information that investigators submit to IRBs is confidential in nature, and unauthorized discussion of that information by members could result in their liability for harm suffered therefrom.

Imposition of personal liability in any of the situations described would depend on the plaintiff's ability to prove that the IRB failed to use ordinary, reasonable care in coming to its decision and that because of that failure, the plaintiff was harmed. If the injury would have occurred regardless of the IRB's action, there can be no liability.

This discussion has focused on the legal history of IRBs, the evolution of federal regulations governing their activities, the considerations of the legal requirements to which IRBs are subject, and the consequences of failure to comply with those requirements. No attempt has been made to address state statutory or regulatory mandates that may impose additional duties and liabilities upon IRBs and IRB members. Since state statutes or regulations may be more stringent than federal regulations, compliance with those is imperative.

Reference Notes

1. *Slater v. Baker v. Stapleton,* 2 Wils 359, 95 Eng. Rep 860 (K.B. 1767).
2. *Slater v. Baker v. Stapleton,* 2 Wils 359, 95 Eng. Rep. 860 (K.B. 1767).
3. *Carpenter v. Blake,* 60 Barb (481) (N.Y. 1871).
4. *Carpenter v. Blake,* 60 Barb (481) (N.Y. 1871).
5. *United States v. Karl Brandt et al.,* Trials of War Criminals before Nuremberg Military Tribunals under Control Council Law No. 10 (October 1946–April 1949.
6. *The Institutional Guide to DHEW Policy on Protection of Human Subjects,* 1971, Department of Health, Education, and Welfare, Washington, D.C.
7. Pub. Law 93-348, The National Research Act.
8. Pub. Law 87-781 87-781; 21 CFR Part 312.
9. *Hyman v. Jewish Chronic Disease Hospital,* 206 N.E. 2d 338 (1965).
10. *Kaimowitz v. Department of Mental Health,* Civil No. 79-1934 AW, Circuit Court for the County of Wayne, State of Michigan, July 10, 1973.
11. *Wyatt v. Stickney,* 344 F. Supp. 373 (M.D. Ala. 1972), 344 F. Supp.
12. 43 *Federal Register* 53950, Nov. 17, 1978.
13. *Wyatt v. Stickney,* 344 F. Supp. 373 (M.D. Ala. 1972).
14. *New York Association for Retarded Children v. Carey,* 393 F. Supp. 715 (E.D. N.Y. 1975).
15. 42 *Federal Register* 3076, Jan. 14, 1977.
16. *The Institutional Guide to DHEW Policy on Protection of Human Subjects,* 1971, Department of Health, Education, and Welfare, Washington, D.C.
17. *Tunkl v. Regents of California,* 60 CAL 2d 92, 383.
18. 45 CFR 46 Subpart C.
19. 45 CFR 46.206–46.210.
20. 43 *Federal Register* 31786, July 21, 1978.
21. 43 *Federal Register* 53950, Nov. 17, 1978.
22. 43 *Federal Register* 56174, Nov. 30, 1978.
23. *Federal Register* , Jan. 26, 1981.
24. 45 CFR 46.107 (a).
25. 45 CFR 46.107 (a).
26. 45 CFR 46.111 (a) (1).
27. 45 CFR 46.112.
28. 45 CFR 26.112.
29. *Board of Regents v. Roth,* 408 U.S. 564 (1972).
30. *Mathews v. Eldridge,* 424 U.S. 319 (1976).
31. *Nielsen v. Regents of the University of California,* Civil No. 665-049 (Superior Court of California, County of San Francisco, Sept. 11, 1973).
32. *Bailey v. Mandel,* Civil Action No. K-74-110 (D.C. M.D. 1974).

References

Gray, B. H., 1977, *Human Subjects in Medical Experimentation,* p. 11, John Wiley, New York.
Hershey and Miller, R. D., 1976, *Human Experimentation and the Law,* p. 8, Aspen Systems, Germantown, Maryland.

3

General Organization of the IRB

MARY KAY RYAN

While the federal government has provided general specifications concerning membership, quorum requirements, and the principles that IRBs must follow in making their decisions, it has remained silent on the organizational structure of the board, which is discussed in this chapter. This discussion complements that on the procedures for administering the board, which may be found in Chapter 7.

Since many hospitals, colleges, and universities have had precursors of IRBs long before the federal regulations appeared, a historical precedent for the appointment process may already be in place (Melmon *et al.,* 1979; Richmond, 1977). If the institution is fortunate, there may already be acceptance of the IRBs' important role by the research and administrative community. Other institutions, however, may be in the process of establishing an IRB. This chapter will discusss models for the selection of IRB members, the leadership responsibilities of the board, the special problems and concerns in recruiting lay or community representatives, the organization of the board for purposes of review, the concerns of specialized institutions such as schools for the handicapped, psychiatric institutions, and skilled nursing facilities, and the role of regional organizations of IRB members.

The Source of IRB Authority

The IRB's source of authority and its credibility within research and administrative communities is reflected in the appointment process. The IRB should be appointed by an individual who has both administrative stature and

MARY KAY RYAN • Long Island Jewish–Hillside Medical Center, New Hyde Park, New York 11042.

credibility with the institution's research community. Since the IRB may be called upon to make difficult and, for some, unpopular decisions, its authority must not be questioned. While part of that authority rests in the stature of individual IRB members among their peers, and another part flows from federal (and, where applicable, state) regulations, a major part of the board's authority and autonomy lies in the appointment process.

Generally speaking, the board should be appointed by the chief executive officer of the institution upon recommendations received from his chief administrators responsible for research (for example, Provost, Vice President of Research, Dean for Academic Affairs, and Dean of the Faculty are commonly used titles). Advice may be sought from the chairmen of departments involved in research with human subjects or in ethical issues; recommendations may be solicited from the research committee or faculty senate. The chairman of the IRB should have a direct reporting relationship to the chief executive officer in order to insure the autonomy of the board. In many institutions, the administrator directly responsible for research may be designated as the first line administrator with whom the IRB chairman may discuss policy issues and specific problems.

In state-wide university systems, where there is usually a more complex organizational structure, concern for the welfare of participants in research must be in evidence at each level. In multiinstitutional state systems, the chancellor/president may appoint a "supra-IRB." This university-wide board is made up of representatives of several colleges within the system, legal counsel, and community representation. This board is not called upon to review proposals, but is established to create uniform policies and standards of review for the entire university system. It can provide a mechanism for appeal for researchers at odds with a local IRB, and can advise the chairman of constituent IRBs on difficult points and on the implementation of new regulations. Its less tangible (but most important) function is to demonstrate concern at the highest institutional level for the protection of human subjects. In such state-wide systems, the chancellor/president typically delegates to the president of each institution the authority to appoint a local IRB.

A two-tier review system may be appropriate to medical schools which are affiliated with several teaching hospitals, each of which may be located in different parts of the state or in different parts of a single city. It is not uncommon for affiliated medical schools to serve different racial and ethnic groups in very different settings through their affiliated hospitals. In such university medical schools, the IRB may conduct a complete review of the project for research being conducted locally but review only the ethical and scientific soundness of projects being conducted at its affiliated teaching hospitals. The IRB located at the affiliated teaching hospital conducts the second tier review, which addresses the specific research program in the context of the subjects to be

"There are no great men, my boy—only great committees."

Cartoon by Chas. Addams © 1975, The New Yorker Magazine, Inc.

recruited, the consent process, and the intelligibility of the consent form to the community from which subjects will be drawn.

When the local institution is in itself a multicollege system (e.g., a university with a single president made up of a college of arts and science, a law school, schools of medicine, dentistry, nursing) the president may choose to appoint more than one IRB for administrative purposes (Brown *et al.*, 1978). It may be unreasonable (and administratively impossible) to expect a single IRB to be capable of dealing with the complexities of medical research proposals and the host of subtle and speculative problems arising from social and behavioral science research.

At most small to medium-sized institutions the president will appoint the board. This direct appointment mechanism, as opposed to election of the IRB by a faculty senate or research committee, may be preferable for reasons of institutional governance, administration, and accountability. It is virtually impossible for members of a group (such as a faculty senate or research committee) to be held individually accountable for the welfare of the institution. The chairman of the IRB must have a direct reporting relationship to the individual who is responsible for the welfare of the institution and for anticipating and preventing any situation that would place the institution in legal or moral jeopardy.

Responsibilities and Selection of IRB Members

Assuming that each IRB member is aware that his or her primary responsibility is the protection of subjects participating in research protocols, there are many responsibilities and tasks members should be informed of before they accept membership on an IRB. These duties should also be kept in mind as appointments are made.

Chairman

The chairman has several responsiblities that are not shared by other members. The chairman must communicate the IRB's policies to the faculty and should establish a schedule for meetings that will insure timely review of projects. Since the chairman is responsible for the conduct of the board's meetings, administrative skill and expertise in maintaining the flow of review, discussion, and decision-making by the IRB are necessary. This expertise is most important in institutions where consensus of opinion has become a requirement prior to board action (Cowan, 1975; Brown *et al.*, 1978; Gray *et al.*, 1978).

The chairman, more than any other member, must be concerned with the orientation and continuing education of IRB members so that the committee can function with a common understanding of institutional policy and ethical issues involved in the review process. The chairman should oversee a rotation plan for members, directly communicate the board's decisions to researchers and institutional administration, respond on behalf of the institution to the federal and state governments when comments are solicited on proposed regulations affecting human subjects research, and report violations of IRB and institutional policy to the chief executive officer or his designee. FDA regulations also require that the chairman report these violations directly to the federal government.

In addition to these duties, the chairman must be aware of the complex interplay of personalities that occurs in groups to insure that each member is participating in the review process. Of particular concern is full participation by lay/community representatives, many of whom could feel overwhelmed by the professionals on the board (Gray *et al.*, 1978).

In order to fulfill the charge to protect human subjects and carry out the required leadership role, the chairman should be selected from the most highly respected members of the research community, and have long enough tenure at the institution to be aware of the total scope of its activities. A review of the literature reveals very little about how chairmen of IRBs are selected. There are, of course, two methods: election of the chairman by the IRB itself or by a faculty senate, or selection by the chief executive officer of the institution based on widely solicited recommendations. It appears that most institutions follow the direct appointment method. Appointment by the chief executive officer establishes a direct reporting relationship on matters of importance, and avoids situations in which pressure may be exerted on the chairman by a specific academic constituency involved in an election process. The term of office should be a minimum of two years: longer terms of offices should be considered in order to provide the IRB with stability and continuity in the development of its policies and procedures.

Other Institutional Members

Each member of the IRB has the shared responsibility of reviewing proposals in a timely manner; adhering to the administrative policies of the board concerning written comments; attendance at meetings; keeping abreast of changes in federal regulations; and discussing and attempting to resolve with the investigator any problems with a proposal or a consent form that would delay action by the board.

The responsibility to interact with the researcher to resolve difficult issues and ethical questions makes it imperative that the institutional members of the board have not only some understanding of ethical principles governing research but also some appreciation for research methodology. While scientific review usually takes place prior to the proposal being submitted to the IRB, questions concerning methodology inevitably arise during the consideration of a protocol. It may be beneficial to both the subjects and the researcher to have the widest possible review of methodology (May, 1975). The faculty member assigned primary responsibility for the review must resolve these as well as problems arising from the consent form. The interaction between the IRB members and the investigators is a critical determinant of how efficiently the board functions and is perceived by the faculty research community as a whole. For this reason it is important to select faculty IRB members from those who themselves have some experience in scholarly investigation, have demonstrated an interest in their colleagues work, and are willing to advise and be helpful.

The institutional members should be selected from faculty in departments involved in human subjects research and ethical studies as well from the administrative staff. In hospitals and medical centers, the institutional members typically include members of the junior and senior research community, a nurse, a patient advocate, a legal representative, a clinical pharmacist, social service staff, as well as nonmedical administrators of clinical areas where patients are recruited as research subjects.

In colleges and universities institutional membership is also drawn from both junior and senior research faculty, and might likely include members of the humanities departments, a representative of the office of the dean of students (since students are likely to be recruited for social, behavioral, and physiological research) and, possibly, a representative of the student body. Depending on the nature of the research carried out at the institution, there may also be legal representation.

If the institution feels that the presence of an attorney is desirable, the IRB should be aware of the dual responsibilities this member may carry. If the attorney is independent of the institution, then, as with all other members, the primary responsibility is to protect subjects participating in research. If the attorney (or law firm with which the attorney is connected) is retained as counsel by the institution, there is, in addition to the primary responsibility, the duty to protect the institution from legal jeopardy.

Both HHS and FDA regulations require that the IRB be composed of members of varying backgrounds. An institution should select IRB members to provide a broad-based interdisciplinary review of each research project. Regulations require that the IRB be composed of at least five members of varying backgrounds. It may not consist entirely of members of one sex, and there must be at least one nonscientist and one member unaffiliated with the institution.

Institutions may appoint voting or nonvoting consultants or members-at-large for specific research areas where broad-based expertise may be lacking or from which too few proposals emanate to warrant a permanent institutional representative. These consultants or members-at-large must be cognizant of IRB policies and procedures and may include a biomedical engineer or a clinical pharmacist.

The requirement for an interdisciplinary review can pose special problems to specialized institutions, which may have limited professional staff from which to draw. These institutions include state psychiatric hospitals, schools for the handicapped, nursing homes, and skilled nursing facilities. Such institutions could broaden membership by approaching local colleges, universities, medical schools, and hospitals with whom they are affiliated. Similarly, local community and health agencies serving special populations (mentally handicapped, elderly, disabled) could provide representatives for an institutional review board. Another alternative to provide an interdisciplinary review for research is to delegate the review to an IRB at a neighboring academic center.

While each member may be called upon to provide specific technical information relevant to a proposal submitted from a colleague in the same field, each must understand that he or she is expected to bring an intelligence and ethical sensibility to the entire IRB review process.

Lay/Community Representatives

The federal requirement for lay representation evolved slowly in the wake of exposure of research abuses of the informed consent process by biomedical researchers. Reviewing the historical documentation can shed some light on exactly what types of laymen legislators had in mind when they approved regulations requiring noninstitutional representation.

The Surgeon General's 1966 policy statement required that review boards consider the laws of the community in which research is conducted, and later in 1969 indicated that the board "be composed of members with a broad background competent to consider questions of community acceptance of research." DHEW's 1974 regulations set forth the requirement for lay representation and further stipulated that a quorum of the committee could not be made up of a single professional or lay group. In the context of the times, however, it is evident that the intent was to bring into the decision-making process representatives of the population likely to be recruited as research subjects.

In order to bring into the system community representatives, institutions may select from local residents from a variety of backgrounds, attempting to develop an ethnic and racial balance reflective of populations likely to be recruited as subjects.

The federal regulations specify that at least one member of the committee be independent, i.e., neither affiliated with the institution nor related to a person who is affiliated with the institution. Many institutions have interpreted this requirement in ways that have allowed individuals not particularly representative of subject populations to become members of the IRB as the "lay" member. The lay member may be a principal or teacher in a local school, an administrator of a social service or health center located in the institution's catchment area, a representative of a community organization with which the institution interacts, a faculty member at a neighboring institution, or a representative of a city, state, or county health or education department. Such lay representatives usually have above average education levels and, by virtue of their professions, may be predisposed to viewing research in general as beneficial to society (Robinson, 1979).

The problems cited most frequently by institutions in their attempt to recruit lay/community representatives revolve around finding laymen who will attend meetings, take the time to read proposals, and participate in meetings. It is too easy for the layman to feel intimidated by the professional. The orientation process should stress the institution's commitment to the participation of community representatives in the review process, as well as the lay IRB members' rights and responsibilities on the committee.

It is not easy to select independent members of the community to serve on an IRB. Many local communities and groups within the community periodically find themselves at odds with their neighboring university, college, or academic medical center. It is crucial that community representatives with purely political agendas not be appointed to an IRB. While selection of lay representatives is always problematic, it does offer the institution a chance to expand the community's awareness of how academic research is conducted and dispel anxieties about research (May, 1975). An important side benefit is the opportunity to acquaint members of the lay community with the investigative endeavors and commitments of the institution.

While lay members cannot be expected to review the scientific aspects of proposed research, they are quite able to review the consent process and the consent form to insure that the process is fair and the consent form intelligible and free of jargon. The layman is on the board to broaden the perspectives of the board on community attitudes and concerns and to call the professional's attention to incomprehensible consent forms. Lay members have a responsibility to participate in meetings and, by reviewing the research from the point of view of a potential subject, to raise any questions concerning the value of the project, the potential risk to subjects, and confidentiality that come to mind. Since many people can be intimidated or lulled into a noncritical frame of mind by professionals, it is important to seek out persons who are confident of their own intellectual processes to serve as representatives of the community.

Delegated IRB Review and Regional Organizations

Delegated IRB Review

Federal regulations do not require that every institution conducting human subjects research establish an IRB; the regulations only require that an IRB review federally funded research involving human subjects. This distinction should not be lost on institutions that may be so small or may become involved in research on such a sporadic basis that establishing a full IRB may be impractical and very costly. These institutions may formally delegate review to a neighboring institution, which may be better equipped to complete the review process.

A system of delegated review would be possible, for example, between an independent skilled nursing facility and a local hospital or medical school whose professional staff may be involved in overseeing its quality of patient care. The skilled nursing facility may not have an interdisciplinary professional staff capable of reviewing the scientific merit and ethical implications of a variety of projects involving geriatric patients. Similarly, a community or technical college, not heavily involved in research, may choose to delegate the IRB review function to the IRB at a neighboring four-year college.

Such delegation of IRB review would avoid excessive administrative effort involved in constituting and keeping up to date a broadly based IRB in a setting that does little research and that may have a limited faculty. Such a procedure has been recommended by the National Commission for small institutions and specialized institutions (such as state psychiatric institutions) that may have research programs but a professional staff limited to a single discipline. Delegated review of research is more likely to take place in areas where regional or local IRBs have formed informal or official IRB organizations.

IRB Organizations

As the number of IRBs has grown and as institutions have come to grips with specific concerns as a result of federal regulations or ethical problems raised in particular areas of research, informal or official organizations of IRB chairmen, members, and/or interested researchers have begun meeting for the purpose of sharing information about how local IRBs function and deal with specific problems. Such local IRB organizations can address the review and administration of student-originated research projects, which often originate on a college campus but may be carried out in a neighboring institution. Many researchers are involved in multicenter studies where a common understanding of IRB policies, procedures, and standards for consent forms between institutions could expedite the review process. These IRB organizations can share the responsibility for providing educational programs to orient new IRB members, the professional research community, and students involved in disciplines that are heavily directed toward human subjects research. In addition, coordinated responses and comment can be made on proposed federal and state regulations affecting human subjects research. Two such regional organizations of IRBs exist in Boston and in the New York metropolitan area on Long Island.

Institutional Resources Available to the IRB

The administrative requirements of the IRB review process, with its schedule of meetings and flow of paperwork documenting the decision-making process, require that the institution provide resources to permit the IRB to function efficiently.

The most important institutional resource (Brown *et al.*, 1979) is an institution's faculty and administrative staff, which are asked to devote time to the IRB. Depending on the level of research activity at an institution, and the number of proposals each IRB member is asked to review, faculty should be released from other committee and/or teaching assignments up to the level required to permit effective participation on the IRB.

The work of the IRB and the need for continuity in creation and application of its policies and procedures, as well as continuity in the review process, require that support staff be identified to assist the chairman in coordinating the review, notifying researchers of board decisions, following up on research as the IRB may require, keeping detailed minutes of meetings, notifying federal agencies to whom proposals may have been submitted of the IRB's decision, and attending to myriad other administrative tasks associated with the IRB function.

Finally, the institution must render its facilities (meeting rooms, parking areas, etc.) available to the members of IRBs that may have to schedule meetings in the evening to accommodate schedules of the lay members and the professional staff. These requirements for personnel, staff, facilities, and recruitment of community representatives, as well as for the rotation and ongoing orientation of IRB members, link the board to the institution's administration. In order for the IRB to function effectively the board must have a clear channel to the highest levels of administration and its authority must flow not just from federal law but from the policy of the institution, as stated by its board of trustees. The visibility of the IRB is also achieved via the selection process for board membership, the caliber of those asked to serve, and the willingness of the members of the IRB to be available as a resource to its research community.

References

Brown, J. H. U., Schoenfeld, L., and Allen, P. W., 1979, The costs of an institutional review board, *J. Med. Educ.* **54**:294–299.

Brown, J. H. U., Schoenfeld, L., and Allen, P. W., 1978, Management of an institutional review board for the protection of human subjects, *SRA Journal* (Society of Research Administrators) **Summer**:5–10.

Cowan, H., 1975, Human experimentation: The review process in practice, *Case West. Reserve Law Rev.* **25**(472):533–563.

Gray, H., Cook, A., and Tannenbaum, A. S., 1978, Research involving human subjects, *Science* **201**:1094–1110.

May, W., 1975, The composition and function of ethical committees, *J. Med. Ethics* **1**:23–29.

Melmon, K., Grossman, M., and Morris, R. C., 1979, Emerging assets and liabilities of a committee on human welfare and experimentation, *N. Engl. J. Med.* **282**(8):427–431.

Richmond, D. E., 1977, Auckland Hospital ethical committee: The first three years, *N. Z. Med. J.* **86**:10–12.

Robinson, J. A., 1979, Ten ways to improve IRBs, *Hastings Center Rep.* **9**(1):29–33.

The Costs of IRB Review

Jeffrey Cohen

Among the recommendations included in its report on IRBs, the National Commission for the Protection of Human Subjects of Biomedical and Behavioral Research (1978) urged that "at least a portion of the funds necessary to support the operation of IRBs be directly provided [by the federal government] rather than reimbursed through the indirect cost mechanism."* Regardless of the fact that this recommendation has never been acted upon, it recognizes that IRB review is a burden on institutions conducting research. Whether or not the cost of IRB review is directly supported by the government or is included in indirect costs, it is very important that institutions be able to calculate, as accurately as possible, the cost of compliance with human subjects regulations.

This chapter will present an analysis of how costs were calculated at one institution (the State University of New York at Albany) and compare these costs with those, similarly calculated, at a sample of other institutions. It is hoped that this information will enable institutions to more accurately calculate their own costs.

SUNY/Albany Costs †

The State University of New York at Albany is a medium-sized university with an enrollment of approximately 10,000 undergraduate and 5000 graduate

---

*National Commission for the Protection of Human Subjects of Biomedical and Behavioral Research, 1978, *Report and Recommendations on Institutional Review Boards,* p. 9, DHEW Publication No. (OS) 78-0009, Washington, D. C.

† A portion of this information was presented at a workshop entitled "The Administration and Cost of an IRB" presented at a meeting of the Public Responsibility in Medicine and Research in Boston, Massachusetts, October, 1979.

Jeffrey Cohen ● Office for Research, State University of New York at Albany, Albany, New York 12222.

students and is not affiliated with a medical school or other health-related facility. The university has established a general assurance with the Department of Health and Human Services (HHS) that does not permit biomedical or new drug research to be reviewed by the IRB. As a result, virtually all of the research involving human subjects conducted at SUNY/Albany is behavioral, social, or educational. The IRB consists of 7 regular members and 7 alternates, also including two noninstitutional members (one regular and one alternate).

The calculation of the costs of IRB review at SUNY/Albany was based on figures obtained from a statistical analysis of the activity of the IRB from June 1, 1978 to May 31, 1979 (Cohen et al., 1980). During that period, the IRB reviewed 278 proposals for a total estimated cost of over $36,000, or $130 per proposal. This total can be broken down into the following two subcosts: approximately $12,000 ($43 per proposal) for the cost of IRB meetings, and approximately $24,000 ($87 per proposal) for the cost of administration. These costs include direct expenditures by the university and the cost to the university of diverting professional time from other duties. In the next section, we will review, in detail, how these costs were calculated and we will present them in such a way that other institutions can perform these calculations based on their own data.

Calculation of Costs

Table I details the calculations and the figures necessary to make those calculations. Of course, prior to attempting to determine the costs of IRB review, an institution must develop data on the annual activity of its IRB. At SUNY/Albany, these data were compiled by hand from the protocols submitted. This proved to be such a time-consuming and tiresome task, we developed a procedure for putting the information on computer and will obtain the necessary data automatically in the future.

IRB Activity

During the year under review, the IRB met weekly when the university was in session and as needed at other times (usually biweekly) for a total of 41 meetings (1b). The average meeting lasted about 1½ hours (1c) and each member spent about one hour preparing for the meeting (1d). The average attendance at each meeting was 4.3 university members (1e) and one community member (1f). These data provide the basis for calculating the cost of IRB meetings.

Table I. Calculating Costs

1. IRB Activity:	
(a) Number of proposals reviewed per year	278
(b) Number of meetings per year	41
(c) Average length of meetings	1.5
(d) Average preparation time for IRB members	1.0
(e) Average institutional members attendance per meeting	4.3
(f) Number of noninstitutional members attending	1
2. Meeting costs:[a]	
(a) Institutional member-hours for meetings	440.75
(b) Average salary level per hour	19.50
(c) Cost of professional time	8,595.00
(d) Honoraria for noninstitutional members per meeting	80.00
(e) Cost for noninstitutional members	3,280.00
(f) Miscellaneous expenses	—
(g) Total meeting cost	11,875.00
(h) Per proposal	43.00
3. Administrative costs:[a]	
(a) Salary of staff assistant	15,000.00
(b) Secretarial salary	1,000.00
(c) Other clerical assistance	2,000.00
(d) Fringe benefits	5,200.00
(e) Supplies, copying, miscellaneous expenses	1,000.00
(f) Total administrative cost	24,200.00
(g) Per proposal	87.00
4. Costs of IRB review:[a]	
(a) Total costs	36,075.00
(b) Per proposal	130.00

[a] Costs in U.S. dollars.

Meeting Costs

In order to determine the number of member-hours of professional time devoted to IRB meetings, the length of meeting (1c) was added to the preparation time (1d), and multiplied by the attendance (1e) and the number of meetings (1b). For SUNY/Albany, these figures yield a total of 440.75 member-hours (2a). The average hourly salary (including fringe benefits) for faculty and administration at SUNY/Albany is $19.50 (2b), resulting in a cost of $8595 for professional time (2c). This figure is probably the most difficult for many institutions to calculate, since salaries vary so much. The most straightforward approach is to use the average salary figure for the institution as a whole.

In addition to faculty and administration members, each IRB meeting was attended by a community member who was given an honorarium of $75 plus travel for each meeting (2d). This honorarium was multiplied by the number of meetings (1b), giving a total cost of $3280 for the community members (2e).

This total was then added to the cost of professional time (2c), yielding a total cost for IRB meetings of $11,875 (2g). Dividing this by the number of proposals reviewed (1a) gives a figure of $43 per proposal (2h).

One might question the rationale for including the salary of faculty and administrators in the cost of IRB review, since their professional responsibilities include service on institutional committees. Although this argument has some merit, it was decided to include these costs since, without the necessity for IRB activity, these professionals would be serving the institution in some other manner. It would seem reasonable that the diverting of professional time is a real cost that the institution must bear. An argument may also be raised concerning the payment of community members. Again, this is professional time (in SUNY/Albany's case, a minister and an attorney) being diverted away from normal duties. The honorarium given to the community members is in recognition of the value of their time.

Administrative Cost

The cost of administration is the largest component of SUNY/Albany's IRB costs. The major element in this cost is $15,000 for salary of a full-time staff assistant (3a) for the IRB. This has been a budgetary item since 1977. This person acts as a liaison between the IRB and the investigators, processes the proposals, prepares the agenda for and records the minutes of IRB meetings, communicates IRB decisions, and maintains IRB records. In addition, clerical assistance is provided by a secretary for 5 hours per week (3b) and a work/study student for 15 hours per week (3c). The total cost for clerical assistance amounts to $3000. Fringe benefits (3d) were calculated for the staff assistant and the professional secretary and came to $5200. Supplies, copying and miscellaneous expenses (3e) came to $1000. Total administrative costs amount to $24,200 (3f). Dividing by the number of proposals (1a) gives a per proposal administrative cost of $87.

The full-time staff assistant for the IRB removed a great deal of the burden from the chairperson of the board. Since dealing with human subjects research constitutes the primary duty of the staff assistant, more time on a more regular basis is devoted to communicating with investigators, monitoring research, and keeping up with the latest issues and regulations concerned with human subjects research.

Total Costs

The combined cost of meetings and administration results in a total cost of $36,075, or $130 per proposal. The two major factors determining this cost are the frequency of IRB meetings and the salary of the full-time staff assistant. These are the two major factors that seem to make the system work as well as it does. By meeting frequently the IRB avoids undue delays that would present an obstacle to research. The presence of the full-time staff assistant provides a resource person who can help insure that human subjects are adequately protected.

Costs at Other Institutions

In order to determine whether these costs are, in fact, typical of IRB costs at most institutions, a questionnaire was designed to elicit information on the costs of IRB review. The questionnaire asked the respondents to provide as much of the information included in Table I as possible. The questionnaire was distributed at the 21st Annual Meeting of the National Council of University Research Administrators (NCURA) in November, 1979 (Cohen and Hedberg, 1981). Approximately 150 questionnaires were distributed and 29 were returned. Of those, ten did not contain sufficient information for a calculation of cost. For the remaining 19 institutions, the costs of IRB review were calculated as described above.

Table II. Survey Results

	Institution									
	1	2	3	4	5	6	7	8	9	10
Universities alone (N = 10)										
Number of proposals	75	70	100	12	25	24	200	250	125	400
Cost of meetings[a]	113	111	51	110	95	101	40	52	88	41
Cost of administration[a]	36	43	53	12	77	71	47	83	109	37
Total cost[a]	149	154	104	122	172	172	87	135	197	78
Universities affiliated with medical schools (N = 6)										
Number of proposals	500	1000	150	135	125	300				
Cost of meetings[a]	51	51	88	131	153	168				
Cost of administration[a]	23	26	16	41	146	32				
Total cost[a]	74	77	104	172	299	200				
Hospitals (N = 3)										
Number of proposals	190	150	575							
Cost of meetings[a]	348	60	173							
Cost of administration[a]	13	18	7							
Total cost[a]	361	78	180							

[a] Costs expressed in U.S. dollars.

Table III. Comparable Data by Type of Institution

	Universities	Universities with medical schools	Hospitals	All
Number of proposals	128	368	305	231
Cost of meetings*	80	107	194	106
Cost of administration*	57	47	13	47
Total cost*	137	154	207	153

*Costs in U.S. dollars.

The data on per-proposal costs are presented in Table II. The range of total costs is from $74 to $361 per proposal, with an average for all institutions of $153 per proposal. Almost two-thirds of the institutions had total per-proposal costs of between $100 and $200. Although the sample of institutions is quite small, it represents a broad cross section of research institutions. Institutions ranged in size from 5000 to over 30,000 students, and there are institutions included in the sample from all sections of the country. Even though the size of the sample is too small to allow statistical analysis, one can assume that the $100–200 range represents fairly typical costs.*

There are some further generalizations that may be drawn, with caution, from the results of this survey. It appears that in many cases the larger the number of proposals reviewed, the lower the per-proposal cost. Aside from the obvious factor of spreading the total cost over a larger number of proposals, this result seems to stem from a number of other pertinent factors. If an institution is going to have a sufficient number of human subjects research projects to warrant the establishment of an IRB, then there are minimum costs that must be met, such as staff assistance and regular meetings. If the number of proposals is low, then the per proposal cost is high. Another factor is that institutions with a large number of proposals to review generally devise procedures that deal with the proposals in efficient ways. These procedures, which include the use of prescreening, initial review by subcommittees, and the use of mail routes, reduce the length and number of expensive, full-committee meetings (see Chapter 6).

It is apparent, based on the data from this survey, that IRB costs vary according to the type of institution (e.g., institutions with medical affiliations seem to have higher costs than those without such affiliations; see Table III). Although this variation may result from too small a sample, particularly in the

---

*The costs at the University of Texas Health Center have been analyzed in a recent article (Brown *et al.*, 1979), and their total cost for about 850 proposals came to $100 per proposal. This figure falls within the range calculated in this survey.

case of hospitals, the differences in cost seem plausible. In order for an IRB to approve biomedical or new-drug research, it must include physicians as members. Since their income is higher than that of academics, the cost of meetings would be higher. In fact, in the sample we analyzed, hourly professional salaries were 50% higher on those boards reviewing biomedical research. In addition, it would be expected that a higher percentage of behavioral and social science research conducted at universities without medical affiliations would be classified as nonrisk research, thereby permitting more expeditious processing. The data supports this, although not as strongly as one might assume. Additional research with a larger sample will be required to substantiate the differences in cost.

Effect of Recently Published Regulations

The costs of IRB review depend directly on the procedures followed by the IRB. Although individual institutions have a great deal of flexibility in determining their own procedures, HHS regulations (45 CFR 46) specify the composition and basic functions of all IRBs. Therefore the costs of IRB review are determined, to a great extent, by these regulations, and any additional revisions would have significant impact on the costs of compliance.

On January 26, 1981, HHS published final regulations amending its basic policy for protecting human subjects. The FDA followed with their final regulations on January 27, 1981. Under these regulations, certain categories of research have been completely exempted from IRB review, and other research could be reviewed by one individual (expedited review). The categories of research involve procedures generally considered to present no risk to subjects. These changes will reduce the number of full-board meetings. This will be particularly true at schools, such as SUNY/Albany, where the majority of the research does not place subjects at risk. On the other hand, if a significant amount of minimal risk research is included under expedited review rather than exempted from review, then a heavier burden will fall on one individual. This may require either allowing a large amount of released time for that individual or hiring someone full time for that purpose, depending on the number of proposals reviewed. For many colleges, this could offset any savings resulting from a reduction in the number of board meetings.

Research exempted altogether from review and research now eligible for expedited review will go a long way toward reducing the professional time and effort (and, hence, the cost) associated with IRB review at colleges and medical institutions. Medical institutions, however, have been burdened by new FDA regulations concerning testing of medical devices. Savings in one area are likely to be offset by these additional responsibilities.

Conclusions

The obvious conclusion which can be drawn from all of this information is that IRB review is a very costly process, and will remain so for some time. If one can project the data from this small sample to the entire population of research institutions in the country, the total cost of protecting human subjects could well be over $18,000,000 annually. Since the total pool of research and development funds is limited, it is highly probable that this large sum of money is being diverted from potential research.

It is clear that the recovery of costs of IRB review through the indirect cost mechanism is at best incomplete. There are two basic factors that create this situation, the process of computing indirect costs and the inaccurate calculation of IRB costs. At most institutions, IRB costs are not isolated as a factor in determining the indirect cost rate. These costs are dispensed throughout various accounts. At SUNY/Albany the administrative costs (primarily salaries and supplies) are included in the budget for the Office for Research. Only a percentage of the budget is recovered through indirect costs, thus only a percentage of the IRB costs are being recovered. The cost of the faculty time devoted to IRB activity is even more difficult to recover, since faculty generally report only an approximate percentage of their time for administrative work. It is impossible to determine if this includes IRB activity.

Since the IRB costs are being only partially recovered, the funds for this activity are being diverted from other needs of the university, such as research and instruction. At institutions receiving little federal support, even less of the IRB costs are being recovered. These institutions can less afford to divert funds from other areas. The burden, therefore, is not being equitably distributed among institutions.

Two implications of this must be considered. First, in order for institutions to recover as much of these costs as possible from indirect costs, they must have the most accurate knowledge possible of what these costs are. The procedures for calculating IRB costs detailed in this chapter may help institutions to acquire direct support from the federal government for IRB costs. This support should be based on the per-proposal costs of IRB review at the given institution, in order to distribute the burden equitably.

Of course, there are more important considerations than cost. No one can put a price on the value of protecting human subjects. Several questions, however, remain. Who is to bear the burden of protecting subjects? Is the protection given by IRB review in accordance with the costs involved? Are there procedures which can reduce the costs of IRB review without reducing the protection of subjects? The answers to these questions are not readily apparent, and a great deal of research and discussion will be required to determine the answers. It is hoped that the information contained in this chapter will help in this process.

ACKNOWLEDGMENTS

The author thanks the National Council of University Research Admin-istrators (NCURA) for its cooperation, and Mrs. Catherine Ortega for her assistance in preparing and distributing the questionnaire referred to in this chapter.

Summary

The costs of IRB review at SUNY/Albany were calculated and deter-mined to be $130 per proposal. The results of a survey of a small, but repre-sentative, sample of other research institutions found that most institutions had costs ranging from $100 to $200 per proposal, with an average of $153 per proposal. Based on the data from the survey, less active IRBs seemed to have higher per-proposal costs than more active IRBs, and IRBs reviewing biomed-ical research seemed to have higher costs than those reviewing primarily behav-ioral and social research. The recent human subjects regulations may result in lower costs, owing to decreased need for full board meetings, but some insti-tutions may face equal or greater costs under the new regulations. In addition, new responsibilities are being added to existing duties of IRBs in medical insti-tutions. Additional research is necessary to determine if the protection given to subjects is in accordance with the high cost of IRB review.

References

Brown, J. H. U., Schoenfeld, L., and Allan, P. W., 1979, The costs of an institutional review board, *J. Med. Educ.* **54**:294–299.
Cohen, J. M., and Hedberg, W. H., 1980, The annual activity of a university IRB, *IRB: A Review of Human Subjects Research* **2**(4):425.

THE REVIEW PROCESS

5

General Principles of IRB Review

ROBERT A. GREENWALD

In the preceding chapters, we have discussed the historical, legal, and ethical background that led to the formation of the IRB system, and we have outlined how such committees should be constituted and financed. In Section II, we discuss in greater detail the principles and mechanisms of IRB review, the preparation and evaluation of informed consent procedures and forms, the procedures applicable to new drugs and medical devices, and the continuing review of approved research. These chapters are directed to a diverse audience—lay members of IRB committees, nonscientific professionals, and administrative staff—and, thus, the discussions will cover a broad range of issues as befits the multidisciplinary nature of IRB review.

The IRB draws its primary mandate from federal regulations. These rules and procedures, to be enumerated as we go along, specify the broad nature of IRB responsibility. In actual practice, each IRB must implement the basic principles within the context of the nature of the institution and its structure. The IRB derives local strength from the importance granted it by the institutional administration, as well as by its own actions to ensure its ongoing credibility. In addition, the local IRB is familiar with the strengths and weaknesses of the investigators working within that institution, as well as with the subject population from which research participants will be drawn (Hendrix, 1977). Hence, a local IRB is obviously best able to evaluate a research proposal meant to be conducted within its walls. In the final analysis, a group of locally appointed individuals must read and study the investigator's submission (such

ROBERT A. GREENWALD • Division of Rhuematology, Department of Medicine, Long Island Jewish–Hillside Medical Center New Hyde Park, New York 11042, and Department of Medicine, State University of New York at Stony Brook, Stony Brook, New York 11794.

as protocol, consent form, and supporting documents) and recommend approval or disapproval of the project, with or without modification(s). The principles and procedures by which such reviewers must be guided are the subject of this and the following two chapters.

Components of IRB Review

The requirements for IRB review are established by federal regulation and can be found in the *Federal Register* (1981), in which they were presented based on findings of the National Commission. In order to receive support from the Department of Health and Human Services (HHS), the institution must give an assurance to the Secretary of HHS that an IRB has reviewed and approved research within its purview, in accordance with HHS regulations (paragraphs 46.101–46.124). The first three requirements deal with the nature of the research and the remainder deal with informed consent. The latter subject will be discussed in detail in Chapter 7. The principles pertaining to research are reiterated here, as follows:

1. Risks to subjects are minimized by using procedures that are consistent with sound research design and which do not unnecessarily expose subjects to risk.
2. Risks to subjects are reasonable in relation to anticipated benefits, if any, to subjects, and the importance of the knowledge that may reasonably be expected to result.
3. Selection of subjects is equitable.

Thus there are three major principles underlying IRB review: scientific merit, subject selection, and risk/benefit ratio. The responsibility of each IRB member to evaluate all three of these factors for each research proposal is the crux of IRB review, and we will discuss each of these in greater detail.

Scientific Merit

One of the most perplexing problems facing IRB members is the relationship between the ethical components of IRB review and the scientific aspects of the study. With the exception of specialized institutions (such as psychiatric hospitals), it can be expected that a substantial number of IRB members reviewing a proposal will be nonexperts in the field of the study, and the consensus among lawyers, clergy, community representatives, etc., is that they generally feel unqualified to pass on the scientific merits of a project. Since many institutions have research committees separate from the IRB, and since most projects are reviewed at many other levels, both within and outside the location where the work will be performed, there is a tendency for IRB members to overlook scientific merit in their review process. In point of fact, it would appear that scientific merit is an unavoidable aspect of proper IRB review.

This requirement that IRBs evaluate research design was explicitly stated in the initial proposed federal regulations governing IRBs that were published August 14, 1979. These proposals stated that the IRB must determine that "the research methods are appropriate to the objectives of the research and the field of study" and that "risks to subjects are minimized by using the safest procedures consistent with sound research design. . . ." These two statements were widely interpreted in research circles to mean that the IRB must evaluate the scientific aspects of the proposed project as part of the risk/benefit equation. As Gray (1975) has pointed out, this concept has often appeared in the literature on the ethics of human experimentation, where it has been widely stated that it is unethical to expose subjects to even minimal risk or discomfort if the research design is so faulty as to preclude the possibility of obtaining valid data. When the final regulations were published in January, 1981, the former of the two statements quoted was deleted, and the latter was retained in a slightly reworded form. Thus HHS "softened" the requirement for IRB review regarding scientific merit. Nevertheless, that proviso remains in effect, and in view of the ethical considerations, which simultaneously weigh in the review, scientific merit appears to still require IRB attention.

The problems facing an IRB that must review research design are substantial and complicated. For one thing, the paperwork that the principal investigator must submit and that the IRB members must digest is increased enormously. A short description of the project and a copy of the proposed consent form clearly will not suffice for review of scientific merit. Background material must now include the rationale for the study, details of its design (including plans for statistical analysis of the data), discussions of previous trials and related studies (including animal work, if relevant, e.g., in drug trials), and descriptions of the investigators' credentials as related to the proposed study. Even if all this material is submitted to the IRB merely as a photocopy of a grant originally prepared for a funding agency, the problems of assembling and assimilation of material are obvious.

Once the IRB has all the necessary data in hand, the problem remains of establishing the best mechanism with which to digest and analyze the material. Cowan, (see Chapter 12) in dealing with clinical trials, discusses this problem in greater detail. For the IRB that does not have sufficient expertise to conduct its own scientific reviews, the options are either to accept without modification the review that has been performed at other levels (e.g., peer review councils for national cooperative trials, pharmaceutical firms, and funding agencies) or to solicit the expertise of outside reviewers hired as consultants to the IRB.

The latter approach has two negative aspects. First, it will mean a substantial increase in the expense of IRB review, plus the introduction of undesirable (and probably unacceptable) time lag factors in the review process. Second, it may have an adverse effect on the investigator himself. It can be argued that in certain settings, the use of outside reviewers may compromise the position of the investigator who originated the project. The investigator has a proprietary interest in his ideas and he may feel that their circulation to outside reviewers, especially when their identity may not be known to him, is undesirable. This would be particularly true if the investigator is proposing a trial project involving a small number of subjects in order to test a proposal prior to preparation of a grant to a major funding agency. If an institution established a policy of sending such proposals to outside reviewers without the express consent of the investigator, it risks violating the rights of the scientist himself. If that be the case, then it behooves the institution to perform the review totally "in house" or under circumstances that insure the protection of the investigator's ideas. Thus, the third avenue of scientific review is to have the evaluation performed simultaneously with that of the IRB. In larger centers, the composition of the IRB can be adjusted so as to ensure adequate representation from all available branches of science. If knowledgeable, experienced, research-oriented scientists are present on the IRB in sufficient numbers and if the investigators are instructed to prepare their submissions such that they are intelligible to scientists, albeit from other fields, most of the larger IRBs can probably cope with scientific review. Alternatively, a separate research committee may exist whose review for scientific merit of the project can be performed simultaneously. Finally, it should be remembered that most institutions require departmental chairmen to "sign off" on research projects within their purview. If the chairman's signature is equated with preliminary review for scientific merit this can be of use to the IRB as well. (It should be noted that Cowan (1975) has found that the departmental chairmen of at least one major teaching institution had neither the knowledge nor the time to perform the functions of IRB review.)

Many research protocols involve statistical analysis of data as a means of validating the findings of the study. Statistical techniques have now become quite sophisticated, and experts can usually analyze an experiment in advance and tell the investigator *how many* subjects will have to be enrolled in the trial in order to assure statistical validity of the results. In the report of the National Commission (1978), it was recommended that "the number of subjects exposed to risk in research should be no larger than required by consideration of scientific soundness." Hence the IRB should examine the number of subjects expected to be enrolled as part of its review of scientific merit, and should look for an assurance that the study will not be carried beyond the point needed to acquire the desired information.

Veatch (1979) has raised another vexing problem that relates to scientific validity, namely how an IRB should handle preliminary data. The problem arises in clinical trials when one therapy is being compared to another by a randomization design; the problem may occasionally arise in such trials when preliminary data indicate that one "arm" of the study is proving superior to the other although satisfactory statistical significance may not have been reached. Providing the potential subject with such preliminary data might clearly jeopardize the scientific validity of the final statistical comparison by violating the randomized allocation of patients, whereas withholding the information breaches the tenets of informed consent. There are a variety of possible solutions to this problem, none of which are satisfactory. The issues before the IRB thus come full circle. If it is unethical to approve a study with faulty research design, is it scientifically valid to conduct a study in which the principles of informed consent may alter the eventual composition of the subject group?

When funding agencies review proposals for scientific merit, the credentials of the investigator who will conduct the study in question are an important and integral part of the review. If the proper scientific implementation of the study depends on the skills and experience of a certain investigator, then this component must be part and parcel of the review for scientific merit. In an analogous manner, the IRB conducting a review, which includes scientific merit, must also consider the competence of the investigator(s). This concept goes back to the Nuremberg and Helsinki codes and is embodied in the current federal regulations on IRB procedures, clearly increasing the burden on the review committee.

IRB evaluation of investigator competence can be broken down into several discrete questions. First, and most readily ascertainable, what are the investigator's academic credentials? Which societies, certifying boards, and academic institutions have recognized the investigator's competence by granting him membership, diplomates, and/or titles? Second, and especially pertinent to community hospitals, what is his reputation within the area where the study will be done? Is he recognized as a specialist, authority, or consultant? In some cases, as discussed later (in Chapter 12) by Cowan, a physician/investigator may find himself in a dual role with conflicting interests between patient welfare and successful pursuit of the project, and the IRB must ensure itself that no such conflicts will abrogate the patient's welfare, or that an appropriate neutral third party will be available to participate in the informed consent procedure.

IRB review of the investigator(s)' credentials must also address itself to the question of who will actually be doing the project. Is the investigator with ultimate responsibility for the project, i.e., the project director or principal investigator, the person who will actually be recruiting the subjects? If not, then the credentials of the co-investigators also come into play. In many medical projects, a doctoral-level nonphysician scientist, e.g., a biochemist or psychologist, may be a prime moving force behind a project, and the IRB must satisfy itself that the medical aspects of such a study will be conducted under the supervision of the physician.

Of equal concern are the many projects performed by students that are likely to pass through a medical center IRB. Most of these are class projects and minor studies rather than serious research that generally involve interviews and/or questionnaires. The interview is a difficult research tool to master, and it cannot always be assumed that a baccalaureate or master's student will be able to obtain meaningful information during a short project. There is a big difference between interviewing consumers in a supermarket and talking to sick, frightened, and/or emotionally disturbed patients. An IRB must assure itself that the preceptor in such projects has matters well in hand and that the subjects will not be placed at risk by participating in minor projects that are approved after cursory review because they have a superficial appearance of triviality. Thorough evaluation of investigator competence is important for both protection of human subjects and control of institutional liability.

Evaluation of all these diverse components of scientific merit is a time-consuming and challenging mandate to an IRB. The review cannot be done properly if it is approached in a casual manner. An IRB of interested, knowledgeable persons, supported by adequate staff, and with sufficient time to conduct such reviews, is obviously required. If the mandate to the IRB is to be properly implemented, the institution must be prepared to make available all the resources required to do a proper job.

Equitable Selection of Subjects

The second requirement of IRB review mandated by federal regulations pertains to equitable selection of subjects for a research project. Historically, this requirement emanates from some of the abuses that occurred in the name of medical research, such as the concentration camp experiments of World War II and the Tuskegee Syphilis Study. In the comments following the publication of this recommendation in the report of the National Commission, it was stated that "the proposed involvement of hospitalized patients, other institutionalized persons, or disproportionate numbers of racial or ethnic minorities or persons of low socioeconomic status should be justified."

This requirement must obviously be interpreted in light of the proposal at hand. Studies that involve diseases to which only one racial or ethnic group are generally susceptible, e.g., sickle cell anemia in blacks, psoriasis in whites, and Tay–Sachs disease in Jews, are clear examples of "legal" exemptions from this stipulation. Similarly, many studies of cancer chemotherapy will revolve primarily around hospitalized patients to the exclusion of ambulatory cases. Such cases should not be a problem to an IRB, at least in terms of patient populations. Of greater potential concern might be a multiinstitutional study involving a university hospital plus a municipal center (or a Veteran's Administration hospital) where clear differences in patient populations can be expected. In this case, the IRB must ensure that subject enrollment will be equitable and that no one segment of the patient population in the total community will bear undue share of the study burden. Of the many requirements that an IRB must enforce, this should probably prove to be the easiest to implement.

Risk/Benefit Ratio

In the final analysis, most IRB evaluations eventually concentrate on analysis of the "risk/benefit ratio" of the project in question. The importance of the comparison of these two items emanates in part from the Federal regulations, which state that the following requirement must be met: "Risks to subjects are reasonable in relation to anticipated benefits to subjects and importance of the knowledge to be gained." The equation of these two factors is expressed in all codes of ethics relating to human experimentation and permeates the literature on this subject.

IRB evaluation of the risks of a research project must take into consideration a variety of possible risks. The risk of physical (or psychological) harm is obviously foremost in consideration. Physical harm can include death (e.g., fatal irregularity of the heart during cardiac catheterization), side effects of drugs (e.g., hair loss, skin rash, and kidney failure), stress reactions (e.g., asking recent widows about their bereavement), and thousands of other possibilities. In a drug study, if the patient's current medications are to be discontinued either temporarily (for a "wash-out" period) or permanently for replacement with a test drug (which might be placebo), this is a risk that must be evaluated. The risks of the alternatives to the procedure under review are also part of the equation. Violation of privacy is another potential risk that must be considered, as is the possibility of adverse financial effects on the subjects.

In a similar manner, the possible benefits of a study are also quite varied. Since many medical projects involve treatment of a disease that has not previously responded adequately to conventional therapy, the major potential benefit of participation in the study would be improvement in the subject's medical condition. In addition, since most projects involve "intensified" medical care with frequent visits to doctors, extra laboratory tests, etc., enhanced exposure to health care is another fringe benefit of participation; if these services are offered at no cost, the patient/subject enjoys the added advantage of decreased cost of medical care. The satisfaction of participating in a study that will enhance our knowledge is also a significant benefit that appeals to many subjects.

Establishing a balance between benefits and risks is not an easy matter. No algorithm or formula can be given that will be universally applicable for all studies. Cowan (1975) has listed some criteria used to assess risk. These include the following: (i) whether the experiment is the very first in a series or one that has been performed somewhere previously; (ii) whether the investigator knows that he or she will at all times retain control of the situation; (iii)

whether the investigator has had experience both in the general field in which
he or she is working and with respect to the procedures to be undertaken; and
(iv) whether safeguards or antidotes are available to counteract an untoward
event that might occur in the course of study. The components of risk/benefit
analysis have been well stated by Cowan (1975) and are reprinted here with
permission of the author:

> The major consideration that attends the evaluation of the benefit of an exper-
> iment is whether the results will be of immediate benefit to the individual subjects
> or whether, instead, the results are intended to benefit society at large. In studies
> that may be termed therapeutic (beneficial to the patient-subject) the review com-
> mittees assess the experimental design of the study and the state of knowledge
> upon which the design is based. For example, in a clinical trial in which the pos-
> sible efficacy of treatment with one drug is compared to no treatment or to an
> alternative drug, the investigator must supply the committee with pertinent back-
> ground information indicating (1) the current mode or therapy for the condition
> in question, (2) the basis (or lack thereof) of the current therapy, and (3) the
> evidence in support of the treatment procedures to be studied. Particularly in sit-
> uations where an agent is to be compared with a placebo, the investigator must
> demonstrate that evidence is currently lacking regarding the utility of any form of
> specific therapy. The committee members, armed with this information, appropri-
> ately referenced, will then often go to the library to familiarize themselves further
> regarding the condition to be studied and the proposed treatment regimens. Their
> main concerns are that subjects in at least one of the study groups will benefit
> appreciably from the treatment, that the subjects in the other groups will be no
> worse off than if they were treated by conventional means, and that the experiment
> is properly designed to yield valid data.
>
> Evaluating the benefits of experiments which are nontherapeutic, but which
> are intended to advance our knowledge generally to the benefit of mankind, is
> infinitely more complex. The crucial issue here is the fundamental conflict between
> the rights of the individual and the rights of society. Although this conflict has
> been debated at length in the context of medical research and human investigation,
> no ready formula exists to guide the members of a review committee. The basis for
> determining the expected benefits of a project that is not directly beneficial to the
> patient-subject derives from the collective medical wisdom and experience of the
> committee members. In a sense, the committees have avoided confrontation with
> such global issues as the individual versus society and have focused more narrowly
> on the specific medical issues that are immediately pertinent to the individual pro-
> posals. The willingness of the investigator himself to undergo the proposed proce-
> dure has not generally been accepted as a measure of the benefit of the procedure
> relative to the risk involved. Such willingness may be a sign of good faith, but it
> is recognized that the investigator may have an obsession with his idea or, at the
> least, substantial emotional and intellectual commitment to it. Hence, he or she
> may not be the best person to assess dispassionately the risks and benefits and his
> or her self-experimentation may be nothing more than a zealous act.
>
> In general, review committees have taken the position that it is justifiable to
> use human subjects for nonbeneficial research if they are satisfied that the project
> is scientifically wise, that the benefits to be gained exceed the risks to any one
> subject, that subjects will be fully apprised of their participation in an experiment,
> and that the participant/subjects will be studied only after giving informed con-

sent. The involvement of human subjects in research is viewed by the committees as a cooperative venture between investigator and subject. Although members of the review committees at CWRU (Case Western Reserve University) have acknowledged the right of individuals, be they normal or patients, to volunteer for hazardous procedures where the risks may exceed the benefits, they have not sanctioned any such studies. They are guided by and have been faithful to the one overriding principle: *primum non nocere*—first of all, do not harm.

The risk/benefit ratio may change during the course of a trial, altering the equation by which the IRB may have granted its approval. The example of preliminary data showing tentative superiority of one form of therapy was given above. The other obvious possibility is that during the course of a clinical trial of one form of therapy, evidence for the superiority of a totally different therapeutic approach becomes available. Quoting again from the report of the National Commission, they recommended, "Subjects should not be excluded from known benefits simply because those benefits were unknown or uncertain at the time the research began." Patients should not be allowed to continue in a research protocol of tenuous benefit if therapy of a superior nature becomes available.

Finally, it should be pointed out that the evaluation of the risk/benefit ratio by the IRB depends heavily on the scope and quality of the information supplied by the investigator in his submission to the committee. The onus is on the investigator to describe in his protocol, in detail, the components of the risk and benefit equation, and to include in the proposed consent form those elements that are germane to the informed consent of the subject. If the investigator provides less than adequate information, he can prejudice the risk/benefit equation that the IRB will attempt to construct. IRB members should not hesitate to go to the library, call in consultants (internal or external), or query the investigator for details about risks of the proposed study. It goes without saying that an adequate evaluation of the risk/benefit ratio can only be performed if the input data is complete and accurate.

Credibility

Levine (1979) emphasized the crucial point that the most important factor in the success of an IRB is its credibility, both within the institution and the surrounding community. Clearly, if the IRB is viewed by the scientific community as a responsible organization, composed largely of knowledgeable peers and informed laymen, efficiently carrying out its mandate, and successfully interfacing with the investigators to assure "smooth sailing" of their proposals, cooperation of the investigators will generally be assured. If the IRB is viewed as an "enemy," a "police force," or an obstructionist group of bureaucrats and paper shufflers, the low level of cooperation will impede both scientific pursuits and proper IRB functioning. To prevent the latter, Levine suggested the following four important considerations: (i) the membership of the IRB must be highly respected within the institution; (ii) the IRB must focus its attention on important things and not allow its efforts to be diluted by trivia; (iii) the IRB must avoid a "double standard" wherein research funded by NIH is accorded one type of review procedure and research funded by other sources is handled separately; and (iv) the IRB must avoid developing an image as a police force, creating an implication that the IRB mistrusts the scientific community.

The authority for the IRB must emanate from the chief executive officer at the institution, and this chief executive officer must be prepared to stand behind the actions of the committee. The IRB should probably be separate from the "research committee," if one exists, but one IRB could serve multiple units within one institution (e.g., medical school, dental school, or hospital). The chairman is the key to its success, and he or she should be selected rather than elected. Sufficient support staff to ensure expeditious review must be provided. The role of the IRB within the institution can be publicized within by issuing a manual covering its procedures, by inviting members of the scientific community to attend meetings as guests, or by holding seminars and teaching days on its procedures and policies. To enhance credibility within the community, an invitation to the science or medical reporter of a local newspaper to attend a meeting and report on IRB activities may be most helpful.

IRB credibility can be expected to be inversely proportional to investigator violation of the principles of informed consent. Several studies (Barber et al., 1973; Gray, 1975) have shown a surprising (and alarming) incidence of investigations being conducted without IRB approval, of subjects being enrolled in research projects without granting consents or even being told that they were participating in research, and of investigators failing to implement the decisions made by the IRB. A creditable IRB, which is backed by strong administration and by an effective monitoring system, is required to prevent such abuses.

If the IRB is viewed within the institution as an aloof final authority to whom no appeal is possible, its credibility will probably suffer. In the event of an adverse IRB decision on a particular project, the question of an appeal by the investigator will arise. Most IRBs conduct their deliberations without the principal investigator being present. In their report, the National Commission (1978) purposefully did not recommend a mechanism for appeal from IRB determinations, since they believed that the IRB should have the final word on ethical matters. On the other hand [see Chapter 3 of this document (National Commission, 1978), which comments on the legal aspects of IRB review] an investigator cannot be denied due process, and if an IRB plans to disapprove or radically modify a protocol, the investigator should be asked to request reconsideration and/or personally appear at a meeting of the IRB. Reatig (1980) gives a number of examples of appeals procedures that may be acceptable as part of an institution's general assurance, and concludes that as long as the procedure does not allow an improperly accredited board the power to override an IRB decision, an appeals process is probably acceptable.

Finally, what is the relationship between the IRB and academic freedom. This issue is also addressed in the report of the National Commission (1978, Chapter 3), where it is pointed out that the requirement for prior approval by an IRB could be construed as a violation of a first amendment right to pursue research. The conclusion is that the regulations pertaining to IRB review address the *manner* in which the research is conducted, not the basis of its ideas, the knowledge sought, or the uses to which it will be put. As stated in the report (National Commission, 1978), "The researcher remains free to investigate the topic, as long as he uses methods that will not harm subjects' interests that the state or institution may validly protect." Within its local sphere on influence, the IRB must continually emphasize to the scientific community that it does not intend to encroach upon academic freedom, subject to the need to protect human subjects from disproportionate risks.

Role of the Lay Reviewer

IRBs are generally a heterogeneous group composed primarily of scientists, other professionals (such as nurses, social workers, clergy, pharmacists, and administrators), and lay persons. The term "lay person" as applied to an IRB committee usually refers to a person selected from the community, often representing the community served by the hospital, who is generally not affiliated with the IRB institution (thus fulfilling a mandated federal requirement). The lay representative(s) are often typical of the research subjects who might participate in a study, and their ability to empathize with the potential subject is obviously a great strength that they bring to the IRB.

The focus of the lay person's review is the consent form. If the consent form is unintelligible, confusing, or overwhelming to the lay reviewer, it will obviously be similarly received by a potential subject. Most researchers, especially in medicine, tend to use technical terms and jargon so often that they may be totally unaware of the degree to which such terms permeate their speech and writing. Despite the investigators' best efforts, many consent forms arrive for IRB review with technical terms in place that require modification or deletion. Many IRBs might find it helpful to prepare a glossary of technical terms with their lay equivalents for distribution to investigators within the institution as a guide to the choice of suitable terminology. As an experienced lay reviewer has pointed out (R. Murcott, personal communication), "The only information sheet that the patient can linger with is the consent form." Presenting a form that the lay reviewer has approved for clarity and lack of complexity is a crucial component of IRB review.

Ghio (1980) has summarized some of the factors pertaining to the role of the lay member of the IRB. She points out that lay members should be chosen on the basis of their interest in health and/or scientific matters rather than friendship with IRB members. Lay reviewers must be prepared to ask many questions when scientific matters or medical terminology require explanation. The lay member can provide valuable insight into concerns about the patient's competence to give consent, the possibility of economic factors affecting decisions to participate in research, the special problems of studies on children, and the expected effect of the project on the patient's quality of life. Lay members of the IRB clearly play a crucial role in the development of fair, understandable consent forms for research.

References

Barber, B., Lally, J. J., Makarushka, J. L., and Sullivan, D., 1973, Research on human subjects: Problems of social control in medical experimentation, Russell Sage Foundation, New York.

Cowan, D. H., 1975, Human experimentation: The review process in practice, *Case West. Reserve Law Rev.* **25**:533–563.

Federal Register 1981, January 26, **46**:8366–8392.

Ghio, J. M., 1980, What is the role of a public member of an IRB? *IRB: A Review of Human Subjects Research* **2**(2):7–9

Gray, B. H., 1975, An assessment of Institutional Review Committees in Human Experimentation, *Med. Care* **13**:318.

Hendrix, T. R., 1977, Local institutional review boards, *J. Med. Educ.* **52**:604.

Levine, R. J., 1979, The evolution of regulations on research with human subjects, in: *The Role and Function of Institutional Review Boards,* Public Responsibility in Medicine and Research, Boston.

National Commission for the Protection of Human Subjects of Biomedical and Behavioral Research, 1978, *Report and Recommendations on Institutional Review Boards,* DHEW Publication No. (OS) 78-0009, Washington, D. C.

Reatig, N., 1980, Can investigators appeal adverse IRB decisions?, *IRB: A Review of Human Subjects Research* **2**(3):8.

Veatch, R. M., 1979, Longitudial studies, sequential design, and grant renewals: What to do with preliminary data, *IRB: A Review of Human Subjects Research* **1**(4):1–2

6

IRB Procedures

Mary Kay Ryan

Compliance with Regulations

The steady evolution of federal regulations from the 1966 Surgeon General's Requirement of Peer Review for Human Subjects Research to the National Research Act of 1974, which created the present day institutional review boards (IRBs), has led to regulations that now specify how the IRB is to function, how long records must be kept, what must appear in required minutes, what constitutes a quorum, which committee members make up a quorum, and what prospective subjects must be told about their participation in research. Regulations now also indicate when an IRB and/or the institution can be disqualified from conducting human subject research and when an institution can be disqualified from recovering federal funds for serious violations of the principles governing human research. All this has led to some disquiet on the part of those involved with IRBs while the Department of Health and Human Services (HHS) and the Food and Drug Administration (FDA) have made it clear that their program of reviewing IRBs for compliance with regulations is neither directed to assessing the quality of IRB review, nor to "second-guess" difficult IRB decisions, but rather to review the *procedural aspects* of the review process.

HHS and FDA have taken substantially different approaches regarding compliance review with their regulations. HHS has chosen to rely on a "scout's honor" approach to compliance in the belief that academic institutions and researchers have too much integrity and too much at stake to violate the ethical principles and procedural guidelines set forth in their regulations. HHS has found that an inquiry to an institution is enough to correct shortcomings and,

---

Mary Kay Ryan • Long Island Jewish–Hillside Medical Center, New Hyde Park, New York 11042.

therefore, HHS has no sanctions against institutions or IRBs. HHS does acknowledge, however, that if illicit research is uncovered it can act to withdraw funding and request that previous funding of an illicit project be returned. The FDA, however, has an entirely different mission and this is reflected in its approach to compliance review. FDA's inspection is a procedural review designed and conducted to determine if an IRB is in compliance with federal regulations and is operating in accordance with its own written procedures. The emphasis is clearly on the procedural aspects of human subjects protection. It is appropriate, therefore, to discuss procedural requirements and to review how other institutions deal with such items as minutes of meetings, meetings via telephone conference call and review of protocols by mail, use of subcommittees and the scope of their authority, researchers' attendance at board meetings, review of research undertaken by undergraduate and graduate students, and the IRB's decision making process (e.g., roll call votes, block voting, approval by consensus), appeal process, and continuing review.

In developing and reevaluating its procedures, institutions can benefit greatly from knowledge of how IRBs at similar institutions have addressed these issues. There is a measure of security in knowing that a number of like institutions may have, for example, chosen to create subcommittees to facilitate review, encouraged or discouraged investigator's attendance at IRB meetings, or required third party observers to the consent process. Unfortunately, there are few sources of such information. Regional IRB organizations provide one way of sharing information, and *IRB: A Review of Human Subjects Research*, published monthly by the Hastings Center Institute of Society, Ethics, and the Life Sciences, is an excellent source of information on government regulations as well as how institutions have approached ethical issues arising from specific protocols. "A Survey of Institutional Review Boards and Research Involving Human Subjects" (Survey Research Center, 1978), which appears in the appendix to the *Report and Recommendations of the National Commission for the Protection of Human Subjects of Biomedical and Behavioral Research*, provides the single most comprehensive review of IRB conduct. Robert J. Levine has prepared a detailed report on the Human Investigation Committee at Yale University Medical School, which also appears in the appendix to the report of the National Commission (Levine, 1976).

History of Compliance Reviews

Several events have led to the FDA's inspection program and a more acute awareness on the part of HHS of its responsibilities in this area. In the early 1970s, FDA inspected IRBs only when there was reason to suspect that there were serious deficiencies in the board's operation. In 1973, the FDA inspected a small sample of IRBs operating mainly at mental institutions, children's hospitals, prisons, and nursing homes. The results, while generally favorable, high-

Table I. *Improvement in Ten IRB Problem Areas*[a]

Problem areas	Percentage of IRBs having problems	
	First inspection	Second inspection
Informed consent forms	56	43
Continuing review	40	20
Written guidelines	40	18
Review of informed consent procedures	36	16
Documentation of committee activities	29	19
Substantive committee minutes	22	11
Inadequate material to review	22	8
Procedures for reporting emergent problems, adverse reactions, and protocol changes for committee review	22	7
Mail review	20	7
Not following written guidelines	19	9

[a] From Department of Health and Human Services, Public Health Service, *IRB Compliance Activity Workshops*, November 7, 1980, FDA, Washington, D.C.

lighted some procedural problems regarding committee structure and/or function. At the same time, FDA began noting significant problems in preclinical and clinical research submitted to it for review. The General Accounting Office audited FDA's activities in the area of human subjects protection, Congress held hearings and more funds were provided for FDA to expand its program of monitoring research. The outcome was FDA's Bioresearch Monitoring Program, which focuses on preclinical laboratories, sponsors and monitors of clinical research, clinical investigators, and IRBs. A more intensive inspection of IRBs began in 1977. One hundred IRBs were inspected by FDA, of which fifty-five had an approved general assurance with HEW (now HHS). Ten problem areas identified as a result of the first inspection showed significant improvement at a second inspection (Table I.). While improvement was most dramatic in the forty-five institutions lacking a DHEW-approved general assurance (Fig. 1), it is noteworthy that even at institutions having an approved general assurance, problems were noted with the continuing review process, documentation of committee activities, review by mail, and failure to follow the institution's own written guidelines.

Filing an Assurance

The first question an institution must answer for itself is whether or not it chooses to file an assurance with HHS. Approximately 650 IRBs at 550

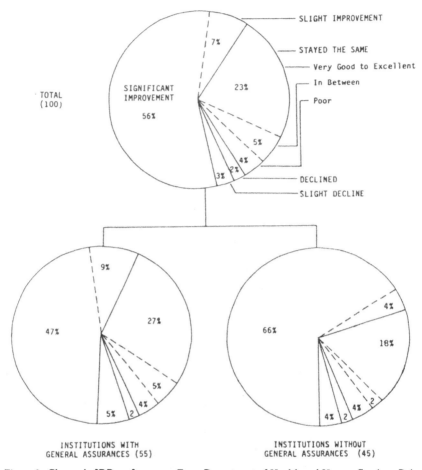

Figure 1. Change in IRB performance. From Department of Health and Human Services, Public Health Service, *IRB Compliance Activity Workshops,* November 7, 1980, FDA, Washington, D.C.

institutions operate under a general assurance.* About 220 of these conduct both HHS- and FDA-regulated research, approximately 3000 IRBs have been approved for review of intraocular lenses alone, and many more IRBs will need to be constituted to comply with the new medical device regulations.†

Prior to July, 1981, an institution filed either a general assurance or a special assurance with HHS. The general assurance described the review and

---

*Department of Health and Human Services, Public Health Service, *IRB Compliance Activity Workshop,* November 7, 1980, p. 54, FDA, Washington, D.C.

†Department of Health and Human Services, Public Health Service, IRB Compliance Activity Workshop, November 7, 1980, pp. 65–66, FDA, Washington, D.C.

implementation procedures applicable to all HHS-supported activities and was required of institutions having a significant number of concurrent HHS projects or activities involving human subjects. A special assurance was required from institutions that did not have an HHS-approved general assurance and described review and implementation procedures applicable to a single project or activity. The final HHS regulations eliminated this distinction and provided a single section (46.103) describing the minimum requirements for assurances. The final rule implements most of the recommendations of the National Commission regarding accreditation of IRB (Survey Research Center, 1978, pp. 10–11) and encourages institutions to have an assurance on file with HHS.

An institution with an approved assurance has sixty days after submission of the application to complete IRB review of the project. Certification of IRB approval is filed on HHS Form 596 (Appendix 7). Institutions not having an approved assurance have only thirty days after an HHS request to certify IRB review and approval of a project. HHS's Office of Protection from Research Risks has prepared a model statement of assurance for use by institutions either filing for the first time or revising their approved assurance as a result of new HHS requirements contained in the final regulations.

Filing an assurance gives an institution an opportunity to examine its institutional policies regarding research. The periodic reexamination of an HHS-approved assurance offers an opportunity for self-evaluation and identification of areas where policies may have altered and procedures modified in the light of experience. In preparing an assurance, care should be taken to reflect upon (i) the institution's overall policies regarding all human subjects research, and (ii) how the IRB will carry out its federally mandated responsibilities. Institutions should keep in mind that their HHS-approved assurance will be the key document in any litigation arising from human subjects research. An assurance should be filled neither with lofty rhetoric nor with procedural minutiae that could easily be modified, overlooked, or deemed unnecessary in the course of time.

Requirements for Assurances

Each institution engaged in HHS-funded research must provide the secretary of HHS with an assurance that it will comply with HHS regulations, and HHS-funded research will be reviewed, approved, and be subject to continuing review by an IRB. It must also (i) provide a statement of principles governing the institution in the discharge of its responsibilities for protecting the rights and welfare of subjects *regardless of the source of funding,* (ii) designate one or more IRBs for which provisions are made for meeting space and sufficient staff to support IRB functions, and (iii) provide a list of IRB members identified by name, earned degrees, representative capacity, indications of experience (such as licenses and board certifications), and employment or other relationships (such as consultant, stock holder, and member of an institution's governing board) between each member and the institution. The assurance must also contain written procedures that the IRB will follow to conduct initial and continuing review of research, to determine which projects require more frequent than an annual review, and ensure prompt reporting to the IRB and to HHS of unanticipated problems.

The assurance is filed with the Office for Protection from Research Risks, National Institutes of Health, Department of Health and Human Services, Bethesda, Maryland 20205. HHS evaluates all assurances in the context of the institution's research activities and the types of subject populations likely to be involved in research, the appropriateness of the initial and continuing review procedures, and the size and complexity of the institution.

The Scope of IRB Review

Since the final HHS regulations make it clear that federal regulations apply only to HHS-funded research, many IRBs may need to reassess the scope of their activities. Research projects may be supported by a wide variety of nonfederal sources: private philanthropy, drug companies, manufacturers of medical devices, scholarly associations, and university or medical center funds specifically earmarked as internal support for research. In addition, research may be undertaken by faculty entirely on their own with minimal institutional resources and by members of the undergraduate and graduate level student body.

One option is for an institution to require IRB review of only HHS-funded research covered by these regulations. The problem with that approach is that it communicates a double standard that will erode the appreciation for standards that should be applied to human subjects research. Deciding to review only HHS-funded research would exempt preliminary and pilot studies usually undertaken before a full-blown application for federal funds is prepared for IRB review. Subjects enrolled in preliminary and pilot studies would then not be afforded the same protection as subjects enrolled in the same study once an application is submitted to HHS. Final reports and, ultimately, publications which emanate from funded research are based on all the data collected during the study; the IRB must ensure that *all* subjects in the trial have given proper informed consent. It seems wiser for an institution to provide a mechanism to review all research involving human subjects. FDA and HHS final regulations exempt certain categories of research from the regulations and permit an expedited review procedure for minimal risk projects. This should greatly alleviate IRB workload.

The scope of IRB review should include review of protocol and consent forms for every project placing subjects at risk. For research approved under expedited review procedures, a summary of the project and an explanation for prospective subjects should be reviewed. Some IRBs choose to have the full protocol and consent document reviewed by every board member. Others provide summaries of the project and consent document to the board with only a subcommittee responsible for the intensive review and recommendations receiving the full description of the protocol and consent document. One suspects that the procedure followed is closely related to the workload of the board and the complexity and technical expertise needed to adequately review projects.

Researchers who do not dispute that their projects place subjects at risk should follow the institutional guidelines and the IRB's policy and procedures in submitting their protocol for full IRB review. Researchers engaged in research falling into an exempted category, or research of no or minimal risk, should provide a brief summary of the project for review in order to obtain concurrence that (i) the explanation of the research to be given to prospective subjects is fair and reflects the protocol, (ii) the research is exempted under the regulations, or (iii) the project actually presents no or minimal risk to subjects. This review can be conducted by IRB staff alone, staff and chairman, staff and another member having expertise in the area of research, or the researcher's departmental committee, if one exists.

Student Research

The FDA and HHS definition of research should exempt a large number of student-originated research projects from the IRB review process. Most research projects undertaken by undergraduates are educational exercises and not research. Many projects undertaken by students enrolled in masters programs fall into the same category. There are, however, certain projects that students undertake which do put subjects at risk. In these cases, it is up to the faculty member in charge of the assignment to review the project and advise the student to seek IRB review. The proposal should be submitted by the student with the endorsement of the faculty member/preceptor.

Students in colleges and universities often seek permission of medical centers, community hospitals, and nursing homes to conduct research projects involving their patients. Some projects may not only be harmless, they may be worthless. The IRB must then assess the degree to which such projects impose on patients, many of whom may be seriously ill and emotionally upset. In any event, each student carrying out a research project in such a setting should submit the project for IRB review (or expedited review) with the name of a faculty member or professional staff member (e.g., Director of Nursing or Director of Social Work) directly responsible for the student while conducting the project. The University of Michigan survey noted that 50% of university IRBs did not review undergraduate research and 20% of IRBs operating at undergraduate colleges did not review student research. Research conducted by graduate students or medical students, however, was reviewed in almost all institutions.

The Decision Making Process

Full IRB or Subcommittee Review

Once an institution and its IRB have formulated overall policies regarding human subjects research, it will face the issue of the most effective and efficient way to conduct the review. The options are to (i) have the full board review everything (full proposal and proposed consent form), (ii) create either standing or ad hoc subcommittees of the board to conduct intensive reviews and to make recommendations to the full board, and (iii) designate a committee member or faculty member as a prime reviewer who will recommend approval or disapproval to the full board. The IRB's choice may be based to a great extent on the number of proposals it must review at each meeting.

Most IRBs followed a procedure of screening out proposals that did not need committee action. These proposals involved use of records and no-risk or low-risk research. Ninety-five percent of the medical schools surveyed assigned the project to an individual for intensive, expert review, but these reviewers were not empowered to make decisions for the IRB. Subcommittees were used by twenty-five percent of the institutions surveyed, and one out of five subcommittees was empowered to make final decisions. The use of subcommittees is most likely to be found in institutions where expert technical opinion is necessary, such as hospitals and medical schools, and where the volume of proposals reviewed at each meeting precludes each member giving each project an intensive analysis from the standpoint of technical soundness and risk or benefit to prospective subjects. For example, Long Island Jewish–Hillside Medical Center has five standing subcommittees: general medicine/surgery; oncology; medical devices; psychopharmacology; and pharmacology, radiology, and investigational drugs. Each subcommittee is composed of a chairman and faculty and lay members who review each protocol. The subcommittee chairman coordinates the review, presents each protocol to the full IRB, and makes the recommendation for approval or disapproval.

The obvious benefit in assigning a proposal to a prime reviewer or subcommittee for initial review is that of expeditious handling. The initial reviewer or subcommittee should resolve any questions before the proposal is placed on the agenda of the full IRB meeting. The subcommittee chairman or prime reviewer presents the protocol to the full board, discusses any questions that were raised,

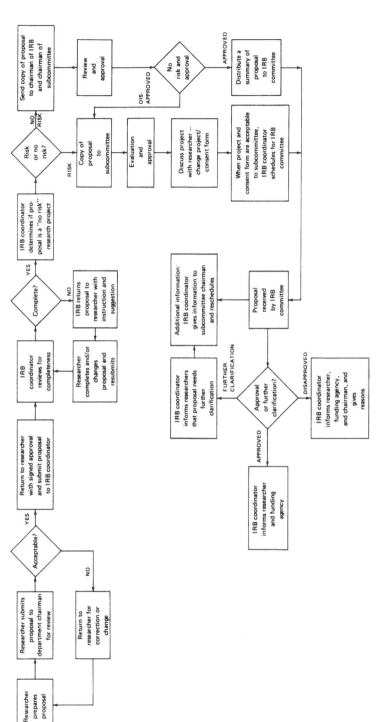

Figure 2. Model of IRB decision making.

discusses the investigator's response, and should be prepared to recommend full board approval or disapproval. The benefits of the subcommittee are weakened if the prime reviewers fail to resolve questions and if they have to turn to the full board for the type of discussion that should take place within the subcommittee. Levine (1976) reported that the IRB at Yale University School of Medicine has a procedural rule that calls for tabling a protocol after its discussion has occupied more than ten minutes. Once the protocol has been tabled, additional information may be requested of the researcher; if necessary, a subcommittee may be formed to reconcile differences of opinion, or additional expert opinion may be sought.

Appeals and Secondary Review

Federal regulations do not cover the question of appeals. However, the preamble to HHS regulations (*Federal Register,* January 26, 1981, p. 8378) makes it clear that if an appeal mechanism is established, the group that hears the appeal and is empowered to make the final decision must be a duly constituted IRB. The preamble states:

> ... investigators do have a right to respond to a negative decision; however, the IRB must finally decide on the ethical acceptability of proposed research involving human subjects. (*Federal Register,* January 26, 1981, p. 8378.)
>
> ... an institution need not conduct or sponsor research that it does not choose to conduct or sponsor, and therefore has final authority to disapprove any research activities approved by the IRB. An institution may not approve research ... which has not been approved by the IRB. (*Federal Register,* January 26, 1981, p. 8380.)

The question of appeals can be a difficult one since mishandling can undermine the credibility and authority of the IRB at its institution. Almost half of the IRBs that took part in the University of Michigan survey permitted appeals. The body that heard the appeal could be the original IRB that had first reviewed the project or other individuals or groups. At universities where there may be more than one IRB, the appeal could be considered by the chairman of another IRB and other representatives of the community, faculty, or student body.

While FDA and HHS regulations do not directly require an appeal process, it is clear that a secondary review by another administrative body may reject an IRB-approved project but may not approve a project that the IRB has rejected. This type of secondary review takes cognizance of an institution not wishing to take on a project for other than ethical considerations. The University of Michigan survey noted that institutions for the mentally infirm were more likely to re-review a proposal, as were other miscellaneous institutions not fitting into the cateogries of universities, medical schools, and hospitals.

The Review Process/Meetings/Voting

The University of Michigan study indicated that two-thirds of the boards surveyed used initial screening. Hospitals and medical schools were very likely to use screening procedures to identify no-risk or minimal-risk research. The screening process should also be directed to reviewing the protocol for conformity with institutional policy. At the very least, each protocol should be endorsed for scientific validity. Many institutions require further endorsement from the chairman of the department in which the research will take place.

IRB staff should conduct the initial screening of each protocol and, if all is in order, assign it for review or refer it to the IRB chairman for review. The researcher should be notified that the protocol has been received and assigned for review and given the name of the primary reviewer in case questions arise. Once the reviewer of subcommittee has completed its review and is ready to recommend action to the IRB, the protocol should be scheduled for discussion at the next meeting. Reviewers should prepare written comments showing the basis of their recommendations to approve or disapprove a study and these comments should be made part of the permanent file.

The University of Michigan survey revealed that the problem most often cited by IRB members was the difficulty in getting together for meetings. The overwhelming majority of boards did, however, conduct business at meetings, while a small percentage used mail review or met via telephone conference calls. Four-fifths of the IRBs surveyed reported that the researcher, at least occasionally, attended board meetings at which proposals were discussed, and over half reported that meetings were open to nonmembers. Whether or not the researchers were present at the vote was not addressed.

There is probably an intrinsic value in conducting the review at a meeting rather than by mail or even telephone. The chance that one thought, question, or comment might spark another question, or disclose a misperception on the part of another member, is more likely to take place in a face to face meeting. The exchange of ideas and opinions serves continually to educate the board member and enhances the IRB's protection of subjects.

Procedurally, a schedule of IRB meetings should be created for the year and should be available to faculty, students, and administrators. Once the review is completed and the discussion concluded, the board should take action by a vote. Voting can be by consensus (found most often in universities); formal voting can be by show of hands, by voice vote (used by the majority of IRBs), or by secret ballot (which is used least frequently). To avoid dominance by any particular group, some boards have chosen "block voting," in which each separate group represented on the IRB—faculty, students, administrators, lay representatives—can cast one vote for their group. This avoids dominance by any particular group. This method of voting is particularly effective where members of one profession (usually the faculty) easily outnumber all other members.

Once the vote has been taken, the IRB should determine the schedule for continuing review. The benefit of assessing the follow-up at the point following decision is that the concerns of board members about the risks involved should be fresh in everyone's mind. Continuing review might be scheduled after a specific number of subjects have been recruited in cases where risk may be serious.

Minutes/Continuing Review

HHS and FDA regulations specify what is to be included in the minutes of these meetings. Minutes of IRB meetings must be kept "in sufficient detail to show attendance at meetings; actions taken by the IRB; the vote on these actions, including the number of members voting for, against, or abstaining; the basis for requiring changes in or disapproving research; and a written summary of controversial issues and their resolution" (section 46.115). Besides detailed minutes, an IRB must keep for at least three years after completion of the research copies of all research protocols; scientific evaluations, if any; approved sample consent forms; progress reports and reports of injuries to subjects; records of continuing review; copies of all correspondence between IRB and investigators; a list of IRB members; and finally statements of significant new findings developed during the course of research that may relate to the subject's willingness to continue participation in the project.

Notification to the Investigator

Following board action, each investigator should be notified in writing of the disposition of his project, the date of the meeting, conditions of approval or reasons for rejection, and, if approved, the schedule for continuing review. Since IRBs tend to modify consent forms, the IRB staff should provide an official copy of the consent form as approved by the board to each investigator.

At least one major university provides a supply of approved consent forms to the investigator, either as needed or up to the limit of the number of subjects which can be recruited before a continuing review is conducted. The majority, however, continue to rely on the investigator to notify the IRB when the designated number of subjects have been recruited and to use the IRB-approved consent form.

Presence of the Investigator at Meetings

As a general rule, if the investigator is not a member of the IRB, he or she is not routinely present when the protocol is discussed. The subcommittee chairman or prime reviewer may ask an investigator to answer questions in projects presenting substantial risk. In the event of controversy, the board may invite the investigator to appear to provide further information. If an IRB member is the investigator or member of the investigating team, a question arises as to how the discussion should take place. To avoid conflict of interest, a board member should not vote on a project in which he or she may have an interest and should be absent when the discussion takes place to permit maximum expression of concerns, if there are any.

Review of Advertising

Once approved by the IRB, many projects may need to advertise to recruit subjects. Since public advertising reflects on the institution, it is appropriate that such advertisements be reviewed by an institutional body. Since the advertisement should accurately reflect the study being performed, IRBs may be asked to review and comment.

Compensating IRB Members

Most IRBs do not pay members. Faculty and administrators who serve on IRBs do so as part of their overall academic and administrative responsibilities. However, the question of payment for lay representatives may arise. While most IRBs do not pay its members, Yale University does compensate its members. Other IRBs reimburse expenses for such items as mileage and parking, and many that meet during evening hours provide dinner.

The issue of compensation is linked to liability coverage for members. Through payments to board members (even a $1 per year), institutions may be able to provide liability coverage for members. The City University of New York, however, negotiated a policy to cover members of IRBs without going through the payment process.

Compensating Subjects

The issue of providing compensation to subjects for injuries caused by participation in research is a difficult one for most institutions. Questions concerning a mechanism to remunerate injured subjects for lost wages, out-of-pocket medical expenses, and so on have been discussed by ethicists and debated by government for several years. In 1977 a DHEW task force recommended that HEW require recipients of grants to provide compensation for side effects. The task force, however, made it clear that no-fault compensation insurance was unavailable from insurance carriers. Investigators point out that harmful incidents are infrequent and there is no evidence that large numbers of participants in research are injured and in need of compensation. Since 1979 government regulations have required that "where research presents risk of physical harm, subjects . . . be advised at the outset whether there will be any financial protection . . ." (Federal Register, Nov. 3, 1978, p. 51559).

It appears that most institutions responded to this regulation with carefully worded statements informing subjects that compensation was not available. Hospitals often indicate that medical facilities are available but usually at some cost to the patient's medical insurer. A presidential commission continues to examine this issue and may recommend that some sort of compensation plan be organized.

The University of Washington in Seattle is noteworthy for its adverse effects compensation program, in effect since 1972. Their no-fault program was covered by commercial insurance for eight years; the plan supplemented the university's general liability insurance. Subsequently, the university, a state institution, developed a self-insurance plan, made possible by the establishment of a special fund in the State of Washington from which claims could be paid. The adverse effects compensation program applies to projects carried out by university personnel under university sponsorship and covers adverse effects resulting from study procedures. The benefits provided are medical expenses directly associated with the adverse effect up to a maximum of $10,000, and such additional expenses compensation as may be agreed upon to by the parties. The Office of Risk Management assumes the responsibility for determining appropriate compensation, arranging payment of the applicable benefits and consultation with the university's IRB, the investigator, his department, and the state's Attorney General.

Participation in Multiple Research Projects

Large research centers may face the problem of monitoring subjects who may join several research projects, either out of interest in the process of research or for the payments some projects offer subjects. There is a greater potential risk in going from a sleep deprivation project to an exercise project, or to a research project on smoking using untested drugs. Not all investigators are aware that in such settings it may be prudent to ask potential subjects about previous research projects in which they may have recently participated. A central registry of subjects could breach confidentiality or be simply impractical. According to federal law, an institution should report earnings in excess of $600 received by any subject in a calendar year. It is probably safe to say that most institutions have not addressed this issue and would not be prepared to make such reports to the IRS.

Conclusion

Whatever procedures an institution choses to implement will be based primarily on the workload facing the IRB, the complexity of research undertaken by the faculty, and the seriousness with which the institution and IRB members take their responsibility. Given these factors, there should be great diversity in how IRBs fulfill their responsibilities. The key to a successful program of human subjects protection lies in simple, clear procedures, free and open dialogue between investigators and members of the board, and detailed guidelines that the faculty can follow. IRB meeting dates should be posted in each department; board meetings should be open to observers; the process for handling appeals should be clear and timely; and investigators should be kept apprised of the status of their proposal throughout the process. The review process must, above all, be timely and responsive to emergency requests. Any system that adds to the spread of bureaucracy in our lives will frustrate investigators, undermine research, and, finally, fail to meet the prime responsibility, the protection of those who participate in the quest for knowledge.

ACKNOWLEDGMENTS

The author acknowledges with appreciation John R. Pettit, Director, Office of Risk Management, and Diana McCann, Director, Human Subjects Office, University of Washington in Seattle for providing information on the university's compensation system.

References

Brown, J. H. U., 1978, Management of an institutional review board for the protection of human subjects, *SRA Journal* (Society of Research Administrators), **Summer**:5.
Department of Health, Education and Welfare, 1977, Secretary's Task Force Report on Compensation of Injured Research Subjects, DHEW Publication No. (OS) 77-003, Government Printing Office, Washington, D. C.
Levine, R. J., 1976, The institutional review board, in: *Report and Recommendations of the National Commission for the Protection of Human Subjects of Biomedical and Behavioral Research* (appendix), DHEW Publication No. (OS) 78-0009, p.p. 4–18.
Survey Research Center, 1978, A survey of institutional review boards and research involving human subjects, in: *Report and Recommendations of the National Commission for the Protection of Human Subjects of Biomedical and Behavioral Research* (Appendix), DHEW Publication No. (OS) 78-0009, pp. 1–87.

Informed Consent

ROBERT A. GREENWALD

No component of IRB functioning attracts more attention or engenders more controversy than the concept of informed consent. For most IRBs, review and revision of the consent form submitted by the investigator constitutes the committee's major interaction with the project. Although the properly functioning IRB must review project design, the criteria for subject selection, the risk/benefit ratio, and similar components of the proposal, most IRBs will reject totally only a small percentage of projects that come up for consideration, and the majority of submitted proposals will eventually be approved, usually subject to some modification of the consent form as submitted by the investigator. Many research proposals originate outside the institution where they will actually be conducted (e.g., clinical trials of new drugs or new procedures sponsored by pharmaceutical firms, or multiinstitutional cooperative therapeutic trials). In most such cases, the research design has usually been constructed with sufficient expertise such that a local IRB is unlikely to find serious shortcomings with the overall proposal; it is generally the consent form and its applicability to the local patient population that requires action by the IRB.

The consent form, of course, is not the same thing as informed consent, as Levine (1979a) and others have stated. Informed consent is the interaction that ensues between investigator and subject when the former attempts to recruit the latter for the project. This interaction generally occurs in a setting where monitoring would be impossible. Should the investigator understate or fail to reveal relevant risks, be coercive in his recruitment procedure, or make unrealistic promises of benefit, the "informed consent" of the subject might still be attainable, and if there were no requirement for a reviewed consent

ROBERT A. GREENWALD • Division of Rheumatology, Department of Medicine, Long Island Jewish–Hillside Medical Center, New Hyde Park, New York 11042, and Department of Medicine, State University of New York at Stony Brook, Stony Brook, New York 11794.

procedure, there would be no constraint on the investigator in pursuit of his project. A simple notation on a patient's hospital chart that he or she "consented" to administration of an investigational drug is clearly untenable in the current climate. Thus the consent form serves as documentation of the informed consent procedure and becomes the crux of the IRB's analysis of the project.

Lebacqz and Levine (1977) have pointed out that the consent form itself serves a different purpose than the act of obtaining informed consent. Whereas the latter is an attempt to provide the subject with a free choice consistent with the best of ethical principles, the consent form documenting this process is primarily designed to protect the investigator and the institution. As these authors have remarked, the consent form tends to give an advantage to the investigator should an adversary proceeding eventually develop. While the presence of a signed consent form within an institution (e.g., on a hospital chart) may be of comfort to the administration and the professional staff, its circulation may also breach the confidentiality of the subject's participation in the project.

Nevertheless, review of consent forms remains the focus of the IRB's considerations and actions. Most investigators submitting to an IRB for the first time may be quite naive about consent form requirements, not only because of lack of prior experience, but also because medical personnel are accustomed to dealing with standard therapeutic and/or diagnostic consent forms as used in general hospital work, rather than research projects. Consents for cardiac catheterization, hernia repair, etc., do not deal with many of the concepts that affect research consent (such as randomized trials, placebos, and alternatives to participation). A literature search under the heading "informed consent" will yield perhaps a few dozen articles written over the past ten years, 80% of which will deal only with clinical consents for such procedures. For example, the widely quoted article by Annas (1978) deals only with clinical consent and does not mention research. Principal investigators therefore often require assistance in composing a suitable consent form for their projects. This chapter provides guidelines for both the investigator preparing his submission and the IRB member who will be conducting the review of the informed consent.

Preparation of Consents

Preparation of the consent form is the responsibility of the principal investigator, not the IRB. As obvious as this may sound, many investigators do not devote much effort to preparation of a suitable consent. This is especially true when a multiinstitutional project is being planned, and a central office prepares a suggested consent that is disseminated to all participating institutions. The local principal investigator often simply forwards the standard consent to his IRB, not realizing that the consent (i) does not comply with local standards, (ii) was designed for a different type of patient population than that dealt with locally (e.g., varying socioeconomic level or educational background), or (iii) omits items required by the local institution. No investigator should accept without critical review a consent form supplied by someone else, even from the sponsoring agency, and no IRB should be willing to accept a sponsor's suggested consent form if it does not meet local standards (regardless of the source—even NIH-sponsored or pharmaceutical consents can be revised).

No person can be said to have volunteered to be a research subject unless he has first understood for what he is volunteering. Any information that might influence him in giving or withholding his consent is vital to an adequate informed consent. Deliberate nondisclosure of facts material to the proposed research amounts to a misrepresentation of that research to the prospective subject. Such deliberate misrepresentation can expose the investigator to a malpractice suit for negligence or to criminal proceeding on a charge of battery. In such situations, an investigator might well be barred from conducting future research at the institution.

It should be the policy of the institution to hold the investigator of record fully responsible for assuring that subjects participating in his study are fully informed in accordance with the terms and conditions of approval of his project and applicable law and regulation. A subject's signature on an approved consent form may not necessarily constitute such assurance. The investigator is expected to have discussed the project with the potential subject fully prior to requesting the patient's signature on the consent form. Where language, educational, and/or cultural differences exist between the investigator and the subject or the medical condition of the subject precludes his informed consent, special precautions should be exercised. Investigators engaged in long-term studies would be well advised to remind the subjects periodically of their participation.

Components of Consent

"Informed consent" means the "knowing" consent of an individual or his legally authorized representative, so situated as to be able to exercise free power of choice without undue inducement by any element of force, fraud, deceit, duress, or other forms of constraint or coercion. The basic elements of informed consent emanate from sound ethical principles and have been reiterated in numerous official documents, e.g., the report of the National Commission (recommendation No. 4, section F) and HEW's subsequent proposed regulations of August 14, 1979. These basic elements can be summarized as follows:

1. A fair and complete explanation of the procedures to be followed and their purposes, including identification of any procedures that are experimental.

2. A complete description of any attendant discomfort and risk reasonably to be expected.
3. A full description of any benefits reasonably to be expected.
4. A complete disclosure of any appropriate alternative procedure that might be advantageous for the subject.
5. An offer to answer inquiries concerning the procedures, risks, benefits, and any matter concerning the research and patient's treatment.
6. An assurance that the person is free to withdraw his consent at any time and to discontinue participation in the project or activity without prejudice to his care or treatment.
7. An instruction that no disclosure of the individual's name or participation in the research will be made.
8. An assurance that the project and the consent form have been reviewed by an IRB (with identification of what an IRB is), and an offer that the subject may contact the IRB (usually through the research grants administrator, whose phone number should be given) if there are any questions or concerns which the subject would like to express.
9. A statement that new information developed during the term of the project will be provided to the subject if that information might affect the subject's continued willingness to participate in the project.
10. An explanation as to whether compensation and/or medical treatment will be available if injury occurs.

The final regulations of January 26, 1981 are in general agreement with these ten components of informed consent, with minor modifications and changes in wording. An admonition against exculpatory language was added, and a statement about the duration of the project was formally added (most investigators had probably already included this information under No. 1 above). Included as item No. 10 in the January 26, 1981 regulations is a vaguely worded phrase, "state the conditions of participation"; nowhere in the accompanying text is the detailed meaning of this phrase given. Further clarification of the intent of this item will have to evolve as the regulations are implemented.

Appended to the list of the ten "required" elements of informed consent were six additional elements to be included where appropriate. These can be enumerated as follows:

1. A statement that the procedure may involve unforseeable risks.
2. A statement on the circumstances for termination of a subject's participation by the investigator.
3. A statement about additional costs to the subject.
4. A description of the consequences of a subject's withdrawal from the study.

5. A statement that significant new findings will be given to the subject.
6. A statement about the number of subjects in the study.

Item No. 5 on this supplementary list appeared as Item No. 9 on the required list of August 14, 1979; it was probably in widespread use for the 18 intervening months and now becomes optional at the choice of the IRB, as indicated. The statement about costs has also been in widespread use and should probably appear on almost every consent form *(vide infra)*. Inclusion of all this information must be weighed against the adverse effect of creating a document so long that the subject cannot absorb it all, thereby inherently impeding his ability to give an informed consent.

Format of Consent

Consent forms for investigational studies may be prepared in either a standardized format or in a free-form style. Most institutions will probably find it expedient to prepare a suggested consent format and make it available to investigators who can then fill in the blanks with the required information. Elaborate and complex projects will probably require a free-form format that must still contain all the elements otherwise required in the standardized format. A blank form in use at the Long Island Jewish–Hillside Medical Center (LIJ-HMC) is shown in Appendix 9. The following instructions are supplied with this consent form to the investigator:

Title of protocol: This line should contain the title of the project as used on all other documents. Care should be taken in cases where two or more protocols with similar names exist; if necessary, the protocol accesion number assigned by the staff should appear on the actual consent if confusion might otherwise result. This line should contain the title in medical and scientific terms, but this is the *only* section on the consent form where such terms are allowed. *The remainder of the consent form must be written in lay language.*

I hereby agree to (have my ward) participate as a subject in the following project: In this space the title of the project should be reworded in lay language. For example, "serial echocardiographic evaluation of left ventricular hypertrophy following aortic valve replacement" becomes "repeated measurements of the size of my heart using a harmless form of sound waves."

I understand the project will include the following experimental procedures: In this space the events which will take place as part of the study should be described. If blood will be drawn, it should be so stated. If tests will be repeated, the schedule should be described. If a drug is to be given, as far as possible, the dosage, length of administration, etc., should be indicated. If the study is double-blind and/or involves placebos, and if patients are to be randomized, this *must* be stated and defined. (The patient must be informed that he or she may not get the active medication if the study is so designed. If the study is short-term and involves a drug not yet available on the market so that supplies will be cut off after conclusion of the trial even though benefit may have occured, this is a risk to the patient that must be stated.)

I understand that the possible discomforts or risks are as follows: In this space state the risks, side effects, etc., which might reasonably be expected to occur from the procedure in question. The "one known case" phenomenon need not be detailed. If a drug is involved, the description of possible side effects should be limited to those that might occur at the doses to be used in the study; the effects of overdoses can be ignored. This section must be clearly understood by a lay person.

I also understand that the possible and desired benefits of this project are: In this space state the benefits which might accrue to the patient from his participation in the project. Benefits to mankind in general, such as better understanding of the disease under study, may also be included.

I am aware that the following alternative procedures could be of benefit to me (my ward): In this space state the other options which are available. In a protocol involving the treatment of a condition, the term "none" is NOT allowed here; in most studies, the alternative is "conventional" therapy, and this should be stated and described. The risks and benefits of the alternatives should also be stated and described.

Compensation Clause

In November 1978, the Department of Health, Education, and Welfare published an unexpected item called an "interim final regulation," which went into effect only two months later, on January 2, 1979. This regulation amended the definition of informed consent to state that subjects in research projects must be advised as to the availability of compensation and medical treatment should injury develop from their participation in the project. As Curran (1979), has pointed out there was no serious debate over the advisability of offering compensation; the problems were that the regulation was vaguely worded and ill-defined, it was promulgated without opportunity for feedback from the research community, and it was put into effect before insurance coverage or other plans could be set into place to provide the arrangements necessary.

In response to this regulation, a variety of compensation clauses were introduced for inclusion into consent forms (Levine, 1979b). Each institution must define for itself what compensation, if any, it wishes to offer. It is not mandatory to offer any compensation, as long as the subject is so informed. If no monetary compensation is to be offered, the following wording (or some variant thereof) is suggested: "The institution will make available hospital facilities and professional attention at ___ (Name of institution or name of an affiliated hospital, should the research study have been conducted at the affiliated hospital) to a patient who may suffer physical injury resulting directly from the research. The expense for hospitalization and professional attention will be borne by the patient. Financial compensation from ____ (the named institution) will not be provided." In the final regulations of 1981, HHS refused to limit this disclosure requirement solely to physical injury.

Financial Risk

If subjects involved in a research project will undergo testing or procedures that they would otherwise not undertake, and if the subjects will be expected to pay for these studies, there is an obvious financial impact on the subject that constitutes a risk and must be mentioned in the consent form. In most cases, the cost of the excess testing is borne by the study's sponsor, and there is little fiscal impact on the subject. In some cases, all medical expenses are borne by the sponsor and there is a net decrease in the cost of medical care for the subject. This is, of course, an inducement in many cases for subject participation, and care must be taken that the financial inducements do not become coercive. At LIJ-HMC, a statement of financial impact is included in each consent form as follows (the investigator picks one of the five phrases for inclusion in his consent form prior to IRB submission): "I understand that in comparison to the costs of medical care which I would normally bear, my participation in this study may (a) greatly increase, (b) increase, (c) have no effect, (d) decrease, or (e) greatly decrease my cost."

Payments to Subjects

If subjects for a project are to be offered cash payments as an inducement to participation, this information must clearly be stated on the consent form. Levine (1979c) has discussed in detail various suggested wordings for payments vs. reimbursements, partial payments, payments for screening procedures, etc. In an ongoing project, as opposed to the simple matter of a $5 or $10 one-time fee for a small blood donation, subjects may become ineligible for continued participation as the project unfolds, may have to drop out for medical reasons, may elect to drop out for their own reasons, etc. A clear statement in the consent form of the payment policy under various contingencies will avoid confusion and misunderstanding.

Additional Clauses

The consent form should offer the subject an opportunity to ask questions about his participation and to contact either an administrative person in the research office and/or a member of the IRB. There should be a statement to the effect that the protocol has been reviewed by the IRB, and there should be a guarantee of preservation of confidentiality. The right of withdrawal without prejudice should be guaranteed as well. The sample consent forms in the Appendix provide examples of wording that may be used.

Minimal Risk, Oral Consent, and Expedited Review

The final regulations of January 26, 1981 substantially narrowed the scope of HHS regulatory activity in many areas of research. Exempted from IRB review were broad categories of educational, behavioral, and social science research that involve little or no risk to subjects. Included in the exemption were projects in educational settings, such as evaluation of instructional strategies, observation of public behavior, and study of publicly available documents, records, or specimens. The exemption in these categories is applicable only insofar as confidentiality is maintained. If the subjects can be identified from the research records, IRB full-committee review would appear to be required. In addition, state and local regulations should be borne in mind when exempting proposals from review.

These regulations also provided a definition of "minimal risk," viz., "that the risks of harm anticipated in the proposed research are no greater, considering probability and magnitude, than those ordinarily encountered in daily life or during the performance of routine physical or psychological examinations or tests." The 1981 regulations provide for "expedited review" of such projects posing no more than minimal risk. The general categories of biomedical research offered such review include studies of body tissues which can be obtained noninvasively (such as excreta, hair, nail clippings, and expelled placentas), recording of data by body-surface sensors, voice recordings, moderate exercise by healthy volunteers, collection of modest amounts of blood by venipuncture, and use of surveys and psychological tests on normal volunteers. The provision for expedited review was provided in order to lessen the workload of the IRB, not to eliminate the need for IRB review or to abrogate the requirement for informed consent. *The fact that a project can be deemed "minimal risk" does not eliminate the need for submission to the IRB, nor does it eliminate the requirement for informed consent.* It merely allows the IRB to use a review procedure that involves less paperwork and faster action (section 46.110). There is nothing in the 1981 regulations that permits any category of human subject research to be conducted without informed consent.

A provision is made, however, for oral consent and for a "short form" consent procedure. In virtually all projects, with certain exceptions (see below), a written consent is required. For most biomedical projects, this will be the regular, sometimes lengthy form described in this chapter. Provision is made, however, for an oral consent to be obtained wherein the investigator must disclose to the subject, in the presence of a witness, all of the elements of informed consent listed above. The subject then signs a "short form" consent attesting to the fact that he or she has heard the investigator's presentation (which must have been summarized in writing and submitted to the IRB for review). The witness must also sign the short form and must further sign a copy of the summary of the presentation. The subject then gets a copy of the short form and of the summary. The logistics of this procedure are obviously rather cumbersome, and the provision for oral consent would appear to have very limited applicability. Its effectiveness as a liability deterrent for the institution or investigator is probably minimal.

There are two provisions in the 1981 regulations for waiver of the requirement for written consent. If the IRB finds that the only record linking the subject to the project would be the consent form and that the principal risk to the subject would be a breach of confidentiality, then written informed consent is not required. Finally, but very importantly, the regulations state that signed consent may be waived if the IRB finds "that the research presents no more than minimal risk of harm to subjects and involves no procedures for which written consent is normally required outside of the research context." This would appear to mean that most of the projects eligible for expedited review, such as obtaining blood in small amounts, no longer require a written consent form. However, an IRB submission is still necessary, and the IRB may require that the investigator provide the subject, in conjunction with the implicit (but not specifically required) oral explanation of the project, a written statement regarding the research. It should also be noted that the requirement for notifying the subject of the availability of compensation for injury applies only to research of greater than minimal risk, implying that this factor need not be considered in expedited review/no-written-consent circumstances. Finally, it should be remembered that the exemptions provided by the January 1981 federal regulations may not override state, institutional, or legal requirements.

Disposition of Consent Forms

The consent form is initiated by the investigator with assistance, as needed, from the research administrative staff. After IRB review and adoption of a final, approved consent form, the investigator can prepare a sufficient number of forms for the anticipated enrollment in this project. Modern technology allows for much sophistication in consent form preparation if the institutional or protocol budget will allow. For example, word processing equipment can be used to generate an individualized consent form for each subject, to be issued by the research office on request of the investigator. This would allow the staff to keep track of enrollment in the project. By using NCR paper, copies of the consents can be prepared so that the research office, the investigator, and the subject could all have them on file (with precautions to preserve anonymity). On a simpler scale, a blank consent form can be signed by the subject, photocopied on the spot so that the subject gets a copy, and filed by the investigator until an audit requires its retrieval. In the latter instance, the research staff cannot directly monitor enrollment in the project and must request this information from the investigator.

It was generally agreed that subjects are entitled to a copy of their completed consent form; this was made mandatory in January 1981. The original should be maintained in the investigator's *research* (i.e., not clinical) file on the study so as to be available in an audit. Sending a completed and signed copy to the research office would breach confidentiality; likewise, all efforts must be made to preserve the anonymity of the subject when dealing with the project sponsor.

Signature of Witness

Many institutions use consent forms that provide space for a witness to sign under the name of the subject. The witness's sole function is to certify that the subject actually signed the form, i.e., that the subject's signature is not a forgery. As of this writing, there is no federal regulation requiring a witness's signature on a regular written consent form, but state laws vary and may contain such a requirement. Since the consent form, in the legal view, is only evidence that a discussion has taken place between the investigator and a potential subject for the purpose of obtaining the latter's informed consent to participate, it can be argued that the signature of a witness is not required (Holder, 1979). In practice, it is generally not difficult to obtain the signature of the subject's spouse, a clerical worker, etc., for whatever degree of instutional and/or personal protection this may afford. The January 1981 federal regulations provided for a short-form written consent document attesting to the fact that the elements of informed consent were presented orally. In such cases of oral consent, a witness is required and must sign the consent form along with the subject.

Comprehensibility

Clearly, a subject cannot be said to have granted informed consent if the consent form was incomprehensible. The use of complex medical language in consent forms is widespread and unacceptable. In one recent study, sixty consent forms used in clinical cancer trials were analyzed on a readability scale and were found to have been written at a "difficult" level, closer to that of a medical journal than to that of lay literature such as magazines or newspapers (Morrow, 1980). Similarly, Cassileth *et al.* (1980) studied patient recall of information from consent forms and found that satisfactory recall was attained only for well-educated patients, since the comprehensibility of the material was poor. Many investigators have little comprehension of the type of wording required for a consent form that is to be understood by a lay person. The participation of lay members of the IRB in the evaluation of consent form language can be most helpful. At larger institutions, it may be worthwhile to prepare a glossary providing lay terms equivalent to commonly used medical phrases. Care must also be taken in the opposite direction, i.e., to avoid being patronizing in dealing with a lay population that has become increasingly sophisticated in its medical knowledge, partly as a result of the influence of popular television shows.

With enough time and effort on the part of the investigators and the IRB, it would appear that even rather complex studies can be made comprehensible to volunteers on research protocols. Woodward (1979) reported on a study conducted at the University of Maryland involving volunteers who were admitted to a research ward for a cholera vaccination project. Lengthy discussions with the subjects, including a slide presentation, were used to explain the disease under study and the nature of the project. Testing the subjects' comprehension with multiple-choice questions revealed a very high level of understanding, greater than that of a group of medical personnel who did not receive the explanations. The test was used as a screening device, and two subjects were excluded from the project when their test scores indicated a low level of comprehension. The author concluded that volunteers with no special level of educational background could be made to assimilate and comprehend a large amount of information if it was properly presented.

Another mechanism for enhancing subject comprehension about research participation is to allow the potential subject to take the consent form home and read it at leisure before enrollment (Morrow et. al., 1978). Subjects allowed to study the form in this manner appear to possess greater information about the project than those who sign immediately after hearing the investigator's presentation. This system is clearly more feasible than that used by Woodward (1979).

Brady (1979) has described a system used at The Johns Hopkins Medical School in which prospective subjects for a residential sociologic study are invited to a series of briefings and familiarization sessions in the actual research setting prior to the consent procedure. During this preexperimental phase, the subjects receive monetary rewards and are given a manual which describes the experimental procedures. The system works very well at Johns Hopkins, but it is obviously expensive and not feasible for most projects. Not only does such a procedure increase the investigator's (and the institution's) confidence in the nature of the consent, but it also reduces the risk of abortive experiments. As in most other endeavors, the yield is proportional to the input. Further details on this project can be found in Chapter 1.

Conclusions

Preparation of an informative, clear, comprehensible, and nondeceptive consent form is one of the research investigator's most important responsibilities. The investigator initiates the consent form, creates it as an adjunct to his experimental design, and hopes that it will reflect the diaglogue that will eventually ensue between himself and the potential subject during and after the enrollment of the latter. There is a widespread tendency for investigators to treat this task with less enthusiasm and seriousness than it deserves, and IRBs must within their institutions stress the magnitude of the responsibility for preparation of a suitable consent form. On the other hand, an IRB should never accept without critical review the consent form submitted by even the most experienced investigator. Vigilant review and editing (as necessary) are equally important concomitants of the generation of suitable informed consent.

References

Annas, G., 1978, Informed consent, *Annu. Rev. Med.* **29**:9.

Brady, J., 1979, A consent form does not informed consent make, *IRB: A Review of Human Subjects Research* 1(7):6.

Cassileth, B. R., Zupkis, R. V., Sutton-Smith, K., and March, V., 1980, Informed consent—Why are its goals goals imperfectly realized? *N. Engl. J. Med.* **302**:896.

Curran, W., 1979, Compensation for injured research subjects, regulation by informed consent, *N. Engl. J. Med.* **301**:648.

Holder, A., 1979, What commitment is made by a witness to a consent form? *IRB A Review of Human Subjects Research* 1(7):7.

Lebacqz, K., and Levine R., 1977, Respect for persons and informed consent to participate in research, *Clin. Res.* **25**:101.

Levine, R., 1979a, Address to the PRIM & R conference, in: *The Role and Function of Institutional Review Boards and the Protection of Human Subjects*, Public Responsibility in Medicine and Research, Boston.

Levine, R., 1979b, Advice on compensation: More responses to DHEW's "interim final regulation," *IRB: A Review of Human Subjects Research* 1(2):5.

Levine, R., 1979c, What should consent forms say about cash payments, *IRB: A Review of Human Subjects Research.* 1(6):7.

Morrow, G., 1980, How readable are subject consent forms? *J. Am. Med. Assoc.* **244**:56.

Morrow, G., Gootnik, J., and Schmale, A., 1978, A simple technique for increasing cancer patients' knowledge of informed consent to treatment, *Cancer* **42**:793.

Woodward, W., 1979, Informed consent of volunteers: A direct measurement of comprehension and retention of information, *Clin. Res.* **27**:248.

Research on Investigational New Drugs

MARY KAY RYAN, LAWRENCE GOLD, AND BRUCE KAY

There is a growing national concern with drugs in our society, including the ways in which drugs are developed, tested, and marketed by pharmaceutical companies, the extent and effectiveness of the Food and Drug Administration's (FDA) control, and the possibility of unforeseen long-range harmful effects to individuals taking certain drugs.

In spite of this concern, however, drug therapy remains the single most powerful medical regimen in use. Former Secretary of Health, Education, and Welfare Joseph H. Califano testified before the Senate in 1978 that American physicians write about 1.5 billion prescriptions annually. That averages seven prescriptions for every man, woman, and child in the country. Many people feel that a visit to their doctor's office is incomplete unless they walk out with a prescription. There are approximately seventy thousand prescription drug products on the market, and the FDA estimates that there are several hundred thousand drugs sold over the counter and available to the public without prescription.*

Although there is an abundance of drugs to choose from, research continues in an effort to discover new drugs to cure or relieve symptoms of disease. Thousands of patients fail to respond to conventional drug therapy, and there are still too many diseases for which there is neither cure nor relief of debilitating symptoms. The FDA is under great pressure to approve new drugs for

---

* *Proposals to Reform Drug Regulation Laws,* Legislative analysis No. 8, 96th Congress, October, 1979, American Enterprise Institute for Public Policy Research, Washington, D.C.

MARY KAY RYAN, LAWRENCE GOLD, AND BRUCE KAY ● Long Island Jewish–Hillside Medical Center, New Hyde Park, New York 11042.

marketing while insisting that it not be rushed in evaluating drug research, since any error could prove devastating to the public. In a single generation we have witnessed, on the one hand, the almost total eradication of diseases such as polio, smallpox, and diptheria, while on the other hand we have been stunned by the side effects of thalidomide and, more recently, diethylstilbestrol (DES).

In reviewing drug research protocols, the IRB must address several crucial issues relating to the protection of those participating in research and should assure itself that the hospital or university sponsoring such research has created an administrative system that will reduce potential risk. In this section we provide an overview of the genesis of FDA regulations controlling the marketing of drugs; describe the manner in which drug use can be administered in a hospital to reinforce and strengthen IRB review; suggest institutional policy and procedures for drug research; and present specific issues and questions that IRB members should ask themselves as they review drug research protocols. These issues involve the selection of control populations, the use of placebos in research, and the participant's right to a continuing supply of a drug if it proves beneficial where other drugs have failed.

Genesis of Drug Laws

A review of federal control of drugs shows government action only after tragedies and revelations of corporate disregard of the public interest. The first major federal food and drug law was passed in 1906 following publication of *The Jungle*, Upton Sinclair's exposé of the meat packing industry and the scandalous conditions prevalent at that time in the Chicago stockyards. The 1906 Food and Drug Law mandated for the first time that foods and drugs had to be "pure." The issue of how safe these "pure" drugs may have been was not a consideration to law makers of that time.

From 1906 through the 1930s the pharmaceutical industry grew tremendously with no federal requirements other than that of the "purity" of substances used in medication. In 1937, 107 people died when the Massengill Drug Company produced an elixir of sulfanilamide which used as its solvent diethylene glycol, a chemical similar to antifreeze.* This tragedy was the result of the manufacturer's failure to test the toxicity of the product before putting it on the market; the event resulted in the passage of the Food, Drug, and Cosmetic Act of 1938, which required the manufacturer to establish the "safety" of a new drug before it could be marketed.

---

*American Medical Association Chemical Laboratory, 1937, Elixir of sulfanilamide—Massengill, *J. Am. Med. Assoc.* **109** (19):1531–1539.

Thalidomide and DES

Between 1938 and 1962 many drug-related safety problems surfaced, culminating in the thalidomide incident. Thalidomide was first marketed in West Germany in 1958. In 1959, the Merrill Pharmaceutical Company began the process of obtaining FDA approval to market this drug in the United States. Suspecting that the product's efficacy was questionable, Dr. Frances Kelsey was able to delay approval of Merrill's application until 1960, when 80 cases of unusual birth defect were linked to the use of this sedative. By 1961, over 300 cases were reported (Kelsey, 1965). By November 1961 the drug was withdrawn from the German market and six months later Merrill withdrew its FDA application.

The thalidomide incident inspired the enactment in 1962 of the Kefauver–Harris Amendment to the Food, Drug, and Cosmetic Act. This amendment added "efficacy" to the previous requirements that drugs must be pure and safe. Efficacy was to be determined by well-controlled, double-blind studies, and the FDA now required that the results of at least two studies be presented as part of the manufacturer's application. The 1962 law also required the informed consent of subjects participating in new drug studies and reports to the FDA of any adverse reactions.

By the attention and perseverence of Dr. Frances Kelsey, who later received the Congressional Medal of Honor, a tragedy of major proportions was narrowly averted. This was not to be the case with several million women who were given the drug diethylstilbestrol (DES) to assist in the maintenance of pregnancy between 1941 and 1971. Unlike thalidomide, which caused an immediately apparent birth defect in offspring, the problems associated with DES did not become evident until 1971, when the daughters of these women began to show a higher than expected incidence of a rare form of vaginal cancer (Herbst et al., 1971).

The DES incident has highlighted the fact that there is no law that calls for surveillance of drugs once they are approved for marketing. The long-range effect of approved drugs is an unknown. The public, with awareness of potential hazards heightened by incidents such as thalidomide and DES, is suspicious of the drug industry and frequently reluctant to take new drugs.

Potential Hazard of New Drugs

Even though the process for testing and bringing new drugs to market is long and complicated (Fig. 1), unfortunately, neither laboratory testing in animals nor extensive clinical trials can be depended upon as a means of predicting the nature or incidence of adverse drug reactions before a drug is released for general distribution. The most scrupulous work in animals will not

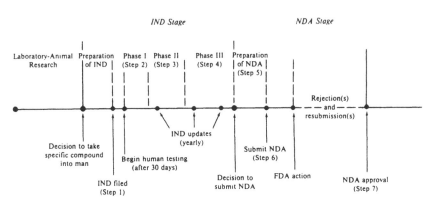

Figure 1. Approval Process for New Chemical Entities. Reprinted from *Proposals to Reform Drug Regulation Laws,* American Enterprise Institute for Public Policy Research, Washington, D.C. © 1979, p. 4.

always predict human toxicity because of species differences or because the effect occurs so rarely that large numbers of animals would be required to detect the toxicity. Similarly, controlled clinical trials, even when carried out on a large scale, may fail to reveal serious adverse effects if such effects occur infrequently. Also, if the population involved in such trials differs substantially from the general population with respect to such factors as age, sex, race, pregnancy, or previous exposure to drugs, side effects may go unnoticed. The rate of hypersensitivity or idiosyncrasy related to the individual's inability to detoxify or metabolize a drug would only be recognized after *tens of thousands* of patients had been administered the drug; obviously too large a population for a clinical trial.

Peculiar and unfamiliar reactions may not be noticed until repeated instances are reported to the FDA Adverse Drug Reaction Reporting Program for analysis and evaluation. Reports from thousands of doctors treating millions of patients may be necessary. Consequently, the first few years of general use of a new drug constitute the "true" clinical trial period. Thus, the evaluation of drug efficacy and safety is not limited to premarketing tests, but is part of the continuing obligation of the practicing physician to his profession and his patients (Lanton, 1965). New drugs are not the only ones that call for postmarketing analysis and evaluation. A drug which has been in use for many years may cause an adverse reaction in rare situations where the patient has unique physical characteristics. Physicians can and have been lulled into a false sense of security about common drugs. Aspirin, for example, was in wide use for over half a century before the medical profession became generally aware of its ability to cause gastrointestinal bleeding (Lasagna, 1968).

This climate of caution and more rigorous attention to drug research is one in which researchers and IRBs must operate in evaluating a range of issues concerning new drugs and new uses for approved drugs. The institution should have as part of its administrative system a comittee structure for the professional evaluation of drugs used commonly in the practice of medicine. The IRB should concern itself with the degree of risk to which a subject may be put as a result of participating in drug research and with how clearly the consent form describes these risks.

Because new drugs are not only given in organized research projects but are also prescribed by physicians treating their patients, reviewing the uses of new drugs is perhaps the area in which the IRB comes into the most direct contact with the clinical practice of medicine. Pharmaceutical companies will make drugs which have not yet been approved by the FDA available to physicians upon submission of FDA Form 1573 (Appendix 5), which states that an IRB has reviewed and approved the drug's use. A physician may request an investigational drug for a single patient, presumably one who has not responded to conventional therapy. Unlike the planned research project where an IRB can schedule review and discussion of the drug and the project at its regularly scheduled meeting, review in clinical situations requires that the IRB and the institution's administrative structure create an efficient, effective, and (above all) nonbureaucratic review procedure that can respond quickly and rationally to immediate needs and emergencies. Anything other than this will frustrate physicians and eventually undermine the IRB review system.

Role of Medical Staff Committees

Before exploring how an IRB can be best constituted to review research and clinical uses of investigational drugs, it is useful to examine the role of medical staff committees in hospitals with respect to drug utilization. The Joint Commission of Hospital Accreditation requires that every hospital have a pharmacy and therapeutics committee; this is responsible for the development and surveillance of all drug utilization policies and practice within the hospital in order to promote rational drug therapy, assure optimum clinical results and minimize potential hazards. This includes the formulation of broad professional policies regarding the evaluation, appraisal, selection, procurement, storage, distribution, use, safety procedures, and all other matters relating to drugs in the hospital. Prior to 1974, when the first federal regulation went into effect, the pharmacy and therapeutics committees of hospitals' medical staff were responsible for policies and procedures to control the use of investigational drugs.

While questions of informed consent and review by community representatives were not considerations in their activities, the professionals who staff these committees were guided by medicine's long tradition of protecting patients and reluctance to experiment on humans. In addition, the Nuremberg Code of Ethics in Medical Research and the Declaration of Helsinki of 1964 (Appendices 1 and 2) provided guidelines for researchers and peer reviewers. National professional organizations also provided ethical guidance. The American Medical Association (AMA) and Pharmaceutical Manufacturers Association (PMA), two major voluntary nonprofit associations, supported strict monitoring guidelines to protect individuals who were otherwise uninformed and without adequate protection.

With the establishment of IRBs under Title 45 of The National Research Act of 1974, the responsibility for setting formalized standards for *investigational drugs* passed from the medical staff's pharmacy and therapeutics committee to the IRB. The pharmacy and therapeutics committee retains major responsibility for drug utilization policies and should act as a resource to the IRB. To assure communication, coordination, and continuity between the two committees, at least one representative should serve on both committees. Often, this individual is the hospital's director of pharmacy services. In this way, the institution can insure a bridge between the independent IRB and the medical staff and administrative committees responsible for drug utilization. In addition, the IRB can become educated about other areas in the institution concerned with drug control. Similarly, the pharmacy and therapeutics committee will have at least one member who can interpret issues and questions with respect to the policy that the IRB review research projects involving drugs as well as certain clinical applications of new drugs.

The National Research Act of 1974 established investigational review boards to review research protocols in the context of informed consent and the risk/benefit ratio to subjects under study. A diverse membership of physicians, allied health professionals, clergy, attorneys, lay people, and community representatives sets the tone for objective review. This constituency is the distinguishing factor when comparing it to a typical medical board committee in which members of the community are not involved.

The remainder of this chapter presents definitions, basic principles, and suggestions for IRBs in reviewing the research and clinical uses of investigational drugs and drugs that, while no longer considered experimental, are being used in innovative ways.

When Is a Drug Investigational?

An investigational new drug (IND) is any drug defined as such by the FDA. Generally these drugs are not available in the marketplace, i.e., in interstate commerce, and they bear the label "Caution: New Drug Limited by Federal Law to Investigational Use." *For purposes of IRB review, an investiga-*

tional drug includes not only those drugs defined as such by the FDA but also any FDA-approved drug used in a nonFDA-approved manner (either by clinical indication, dose, or route of administration as defined by the official FDA-approved labeling that appears as the package insert). In addition, drugs that are not investigational in themselves can be prescribed for a combined effect and the experimental nature of their use lies in the combination. In such cases, these drugs should also come under the definition of nonstandard use and should be reviewed by the IRB.

The use of an FDA-approved drug in a non-FDA-approved manner, either by clinical indication, dose, or route of administration, often poses difficulties for clinical investigators and IRBs. Examples include drugs approved for administration by injection which are administered orally, and *vice versa*, and recommended doses that are exceeded. Even the most common medications can be included, for example, aspirin can be prescribed to cardiac patients to prevent or alleviate clinical symptoms other than its commonly accepted antiinflammatory uses based on its ability to inhibit platelet aggregation.

The issue of nonstandard uses of drugs is complicated by the fact that over the years physicians have developed new uses for old drugs, creating much controversy over the right of the physician to utilize a drug in ways other than were originally approved. It has been our experience that some practitioners strenuously object to obtaining "consent" of patients for an FDA-approved drug, although it may not be prescribed for the approved indication. The FDA has attempted to control the independent utilization of such innovative drug therapy by the "Notice of Claimed Investigational Exemption"* for a new drug. It was the FDA's intention that each physician using an approved drug in an innovative manner would complete the notice voluntarily. It is not, however, a violation of FDA law for a physician to prescribe an approved drug for an unapproved use. Alexander M. Schmidt, a former Commissioner of the Food and Drug Administration, stated that "good medical practice and patient interest require that physicians be free to use drugs according to their best knowledge and judgment. The physician is responsible for making this final judgment, but the law does not require him/her to file an investigational new drug plan. The physician would be advised in all instances when prescribing approved medication for a nonapproved use to file a plan with the FDA" (Luy, 1976). Physicians, however, overwhelmingly ignore this advise.

The FDA's official position was stated as follows in the *Federal Register* of April 7, 1975:

> The labeling (package insert) is not intended either to preclude the physician's use of his best judgment in the interest of the patient or to impose liability if he does not follow the package insert. (It is clearly recognized) that the labeling of a marketed drug does not always contain all of the most current information available to physicians relating to the proper use of the drug in good medical practice.

* *Federal Register*, January 27, 1981, p. 8953.

If an individual is injured owing to the unapproved use of an approved drug, however, the physician who prescribed it, the pharmacist who dispensed it, the nurse who administered it, and the institution where this all took place could be defendants in a law suit. While the official labeling does not conclusively establish the standard of care, the majority of cases seem to indicate that those who deviate from the package insert may be called upon to justify their actions should an injury result.* When formal protocols are submitted to an IRB, the IRB should review the nonstandard uses of drugs, and should be concerned that the subject/patient is informed that the specific drug, while not in itself investigational or "experimental," is being prescribed in a "nonstandard" manner and is forewarned about potential risks. The definition of "nonstandard" use should be applied strictly and permission to allow such use should be based on an expeditious review of recent medical literature supporting the nonstandard use.

In some cases, use of the package insert as the reference point in informing patients and subjects about potential risk does little to resolve any questions. Doctors correctly point out that the *Physicians' Desk Reference*, which lists the FDA-approved package inserts prepared by the drug companies, often fails to mention the most common medically accepted uses for certain drugs, and that it can take many years for these commonly accepted uses to work their way into such references. For example, lidocaine was used for years to treat cardiac arrhythmias (i.e., irregular heartbeat) before this indication was included in the package insert. The package insert, which was developed to give the consumer/patient more direct information about medication, may instead confuse the patient and create needless anxiety about the drug.

We suggest that a proper administrative procedure to permit the utilization of an FDA-approved drug in a non-FDA-approved *but medically accepted* manner requires only the approval of the chairman of the department, a bibliography of medical literature supporting its use, and the approval of the hospital or university administration. The definition of what constitutes a "medically accepted" use should rest on the support of a significant body of scientific literature. Since the use of the drug, while nonstandard, is well documented in the literature, the patient need not sign a consent form, although the patient *should be advised* of the nonstandard use of the medication.

The IRB should be called into the review procedure when an FDA-approved drug is being used in an innovative or experimental manner, support for which is *not* evident in recent medical literature. In such cases—which frequently arise in emergencies or when every conventional therapy has been tried to no avail—the physician should prepare a protocol, obtain the approval of the chairman of his department or his designee, and prepare a consent form for review by the IRB. The use of a drug is considered "innovative" when there

---

* *Mulder v. Parke Davis and Company* 181 N.W. 2nd 880, 1970; *Darling v. Charleston Community Hospital*, 211 N.E. 2nd 253, cert. den., 383 U.S. 946, 1965.

is either no medical literature to support its use for a particular indication or when, in the judgment of professionals, the body of literature supporting its use is neither significant nor conclusive.

Sources of Investigational Drugs

Drugs that are designated as investigational by the FDA are subject to federal controls and can be obtained from the following sources:

1. *The federal government:* Various agencies within the Department of Health and Human Services (DHHS) will make available, through cooperative clinical trials, investigational drugs for a specific use. Typically, such studies are launched either at a major clinical research center or at the National Institutes of Health in Bethesda, Maryland. The base institution supplies the drugs and the recipient institution provides the clinical data on its use.
2. *Regional resource centers:* The federal government has established for antineoplastic agents a network of national multidisciplinary oncology groups based at various medical facilities throughout the country. These groups use a variety of antineoplastic agents developed by the National Cancer Institute. Member institutions that elect to participate in such protocols receive the drugs involved from the designated regional oncology groups and provide the base institution with information on the results obtained from their use.
3. *Pharmaceutical companies:* A pharmaceutical company will make investigational drugs available to physicians depending on the adequacy of the facility and the expertise of the physician-investigator. A physician wishing to obtain an investigational drug from pharmaceutical companies may do so in one of two ways: the physician may obtain the research protocol that the pharmaceutical company has developed and implement it, or may choose to use the drug in an independent investigation for which he must prepare a protocol. In the latter case, appropriate documents (FD 1573, Appendix 5) and the investigator's curriculum vitae must be filed with the FDA, indicating that the institution is acting as sponsor and the IRB has reviewed and approved the research protocol.

Suggested Administrative Procedures for Investigational Drug Use

In hospitals, investigational drugs are used in three situations: (i) research, (ii) nonemergency clinical situations, and (iii) emergency-life-threatening circumstances. In all three cases the IRB is responsible for review. In the first two circumstances—research and nonemergency clinical situations—the IRB reviews the protocol and consent form prior to the drug's use and in emergency situations, the review is retrospective.

The institution's administrative procedures should make clear exactly who may administer an investigational drug. Most hospitals designate members of the full-time medical staff and members of the voluntary medical staff. In teaching hospitals postgraduates in their second and third years or fellows may be authorized to administer investigational drugs.

Registered nurses may administer an investigational drug providing that they have *detailed written* information concerning the drug, and have been given the privilege to give it by appropriate administrative and/or medical authorities. A nurse who administers any drug should be aware of all the effects that the drug might produce upon administration. Therefore, the nurse must be provided with written information regarding the pharmacology (particularly adverse effects), method of dose preparation and administration, precautions to be taken, authorized prescriber(s), patient monitoring guidelines and any other material pertinent to the safe and proper use of the drug. To accomplish this, the physician/research is requested to complete an "Investigational Drug Fact Sheet" (Appendix 12). After review by pharmacy personnel, a copy of the fact sheet is inserted in each nursing unit's research protocol manual, thereby serving as a ready reference source when the investigational drug is ordered for a patient on the unit.

1. *In research protocols* the physician/investigator should apply to the IRB for approval to use human subjects in research. The protocol should be approved by the department chairman or his designee and the IRB prior to implementation. The protocol should be reviewed internally for scientific merit prior to IRB review.

   In some institutions it is the responsibility of the chairman of the department to attest to the scientific merit of all research conducted within a department. Other institutions maintain a standing committee (which may be called the research committee, scientific advisory committee, etc.) to review research projects for their scientific merit. Obviously, there can be no justification for putting a subject at risk in a protocol whose basic methodology or evaluation procedure is so flawed that the data collected would be useless.

2. *In nonemergency clinical situations* a protocol should be prepared as well as a consent form for the patient/subject or his legally authorized representative. The protocol, consent form, and completed "Request for Investigational Drug" (Appendix 11) should be submitted to the department chairman or his designee for approval. Once approved it should be submitted to the *IRB for review by a specific subcommittee*

composed of the chairman of the review board, the pharmacist, a lay representative, and other members as needed. It should be noted that the IRB must establish a mechanism for expeditious review for the clinical use of investigational drugs. The action taken by the review board's subcommittee should be presented to the full committee for retrospective review. Each review board member should receive the protocol and consent form.

3. *In emergency/life-threatening circumstances,* prior approval by the IRB for the drug's use can be waived, but the physician should submit a protocol, an informed consent, and "Request for Investigational Drug" to the chairman of the department or designee for administrative approval prior to utilization of the drug. A situation may arise in which a patient may be comatose or otherwise incapable of giving informed consent, and in which the only alternative to certain death is an investigational drug. If next of kin is unavailable for issuance of an informed consent, it should be the policy of the IRB and the hospital administration that the proposed lifesaving, investigational procedure may be implemented without proper consent. *Notification of the desire to implement the procedure and approval by the department chairman or his designee and the administrator-on-duty at the hospital should be required.*

"Automatic Stop" Orders

There should be an "automatic stop" on orders for investigational drugs that would require that the order be rewritten every 72 hours after the specific time stated on the original order unless (i) the order indicates the exact number of doses to be administered and (ii) the order specifies the exact period of time the drug is to be administered.

General Administrative Procedures

The mechanism whereby a physician orders and obtains an investigational drug from the hospital pharmacy should be part of institutional policy and IRB guidelines.

For Study and/or Treatment of Inpatients In research and nonemergency clinical situations, the staff coordinator of the IRB will provide the director of pharmacy with a copy of the approved protocol and informed consent, which will be maintained on file in the pharmacy. For each patient's initial order, the investigator must submit to the pharmacy a "Request for Investigational Drug" before the pharmacist will dispense the IND.

In emergency/life-threatening circumstances, before the pharmacist will dispense the IND, the investigator must submit to the pharmacy a "Request for Investigational Drug" completed in full and signed by the chairman of the department or his designee; the investigator must also submit a copy of the informed consent and protocol which consists of a description of the drug's use and bibliographic support. Upon receipt of the completed documentation, the pharmacist may dispense the IND. The review board's staff coordinator should be informed of the request as soon as possible.

For Study and/or Treatment of Clinic Outpatients A protocol, consent form, approval by the chairman of the department in charge of the clinic, and "Request for Investigational Drug" should be sent to the director of pharmacy, who will arrange for the drug to be placed in the outpatient pharmacy.

For Study and/or Treatment of Private Patients In studies involving private patients both on and off site, the physician/investigator may retain the option of receiving, distributing, and maintaining inventory records of the drug in the area where the study is conducted. In such a circumstance, the physician/investigator has the responsibility for accounting to the supplier for the disposition of the drug. This option is not to be construed as an exemption from the regulations stated above concerning approval mechanism or allowable sources of INDs.

In all situations the following principles should apply:

1. It is necessary that the pharmacy be notified in writing by the investigator, with the department chairman's countersignature, when the investigator/physician authorizes other physicians to use the investigational drug. Any other requests for use beyond the approval protocol will be referred to the IRB.

2. The pharmacy, upon request of the physician/investigator, will return the unused inventory of the medication to the manufacturer or physician. It is understood that it is the physician/investigator's responsibility to file a drug experience report with the FDA when FDA-certified investigational drugs have been used.

3. *When ordering an investigational drug from the federal government, a government-approved source, or a pharmaceutical company,* the drug may be delivered to the pharmacy or, when the pharmacy is not open, to the security department. The physician should give a telephone number at which he may be reached when the drug arrives. The drug will be delivered to the patient's floor, where the physician must sign, acknowledging receipt of the drug.

4. In situations involving one-time use of INDs (emergency approval and/or one-patient study), retrospective review and approval must be obtained from the IRB. The physician/investigator should submit the

protocol and consent form and evidence of the chairman's approval to the review board.

The Role of the Pharmacist

In 1957, to ensure proper control of investigational drugs in hospitals, the American Hospital Association and the American Society of Hospital Pharmacists (ASHP) adopted a *Statement of Principles Involved in the Use of Investigational Drugs in Hospitals.* In 1962, this statement was endorsed by the American Nurses' Association and later augmented by the ASHP.

Since 46% of biomedical studies involve investigational drugs, chemicals, or blood products (Donehew *et al.*, 1979), the major role of the pharmacist and a hospital's pharmacy department is evident. The pharmacy is the appropriate area for receiving, storing, labeling, dispensing, and disseminating information concerning investigational drugs, as it is for all other drugs. By following the ASHP principles and the recommendations of some excellent articles found in the literature (Donehew *et al.*, 1979, Arbit, 1978; Mandl *et al.*, 1976; Hassan, 1965), the pharmacist can develop policies and procedures that will assure the control and accountability of investigational drugs within the institution. The American Hospital Association should also be consulted.*

The hospital pharmacist should be a member of the IRB and should assist in the monitoring of investigational drug therapy. The literature is replete with articles citing the role of the pharmacist as a member of the IRB, in monitoring investigational therapy, and in staff and patient education programs (Kleinman *et al.*, 1974; Kleinman and Tangera, 1974; Stephen and McKinley, 1974; Hiranaka and Gallelli, 1979). Because of his knowledge of pharmacology, the pharmacist is often the first individual who recognizes the use of an FDA-approved drug for a non-FDA-approved indication, dose, or route of administration. Because physicians/researchers usually practice in several facilities, the pharmacist is usually the first individual to discover that an investigational drug has been introduced from another hospital without the IRB's approval. From this strategic standpoint, the pharmacist can monitor compliance with the institutional policy and the IRB's policies and procedures, thereby assuring the rational and safe use of investigational drugs.

Each institution should have a statement of policy detailing (i) exactly how investigational drugs are to be ordered and delivered to the hospital; (ii) what is required before an investigational drug will be dispensed by the pharmacy (at minimum evidence of IRB approval); (iii) information on precisely

---

* *Reference Manual on Hospital Pharmacy,* pp. 127–131, 178–179, American Hospital Association, Chicago, 1970.

who will administer the drug; and (iv) additional information for distribution to professional staff on dosage, adverse effects, contraindications, route of administration, etc.

Conclusion

Since the enactment of the Food, Drug, and Cosmetic Act, the country has not experienced tragedies on the scale of the elixir of sulfanilamide episode and has managed to limit the impact of thalidomide in the United States. On the negative side, however, there has been a marked decrease in the development of new drugs in this country. While IRBs are not called upon to assess the impact of federal regulations on the nation's drug industry, it should be aware of the impact of overly restrictive regulations on research and development of new therapies and the implications for improvements in patient care.

In the area of new drug development, the first step is the development of new chemical entities (NCEs), compounds not previously marketed that are found in nearly all therapeutic advances. Since 1962, when the Kefauver amendments to the Food, Drug, and Cosmetic Act were enacted, there has been a steady decline in development of NCEs (see Fig. 2) (Kaganov, 1980). Between 1957 and 1961, the average number of NCEs introduced annually stood at 56; since 1962, the number has dropped to 17. There has been a marked decline in the number of companies engaged in research since 1962, and a corresponding increase in the number of new drugs available first to patients abroad (see Fig. 3).

IRBs must be aware of changing regulations in the area of drug research and new proposals emanating from the FDA. Recently, the FDA announced

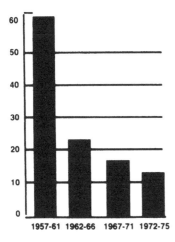

Figure 2. Average annual number of new chemical entities approved by the Food and Drug Administration. Reprinted from Kaganov, (1980).

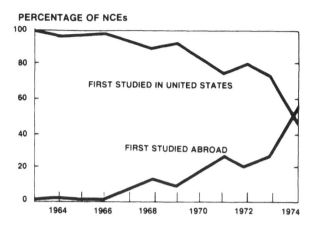

Figure 3. Relationship between the number of new chemical entities (NCE) studied first in the United States and those studied abroad from 1963 to 1974. Reprinted from Lasagna and Wardell (1975).

that new regulations permitting certain responsibilities of early IND research to be directly assured by the IRB would be forthcoming in 1982. These regulations, the FDA hopes, will streamline the review process. However, since they appear modeled on the most recently promulgated FDA medical device regulations, they are more likely to place increased responsibilities for initial judgment and recordkeeping on IRBs.

Within the context of concern for patient welfare, IRBs, physicians, nurses, and hospital administration share an interest in furthering safe development of new therapies for thousands of patients for whom existing medications provide no relief from disease and disability.

References

Arbit, H. M., 1978, Regulatory aspects of investigational new drugs, *Am. J. Hosp. Pharm.* **35:**81–85.

Donehew, G. and Schaumberg, J. P., 1979, The pharmacist's role as a member of the institutional review board, *Hosp. Pharm.* **14:**509–519.

Hassan, W. E., Jr., 1965, *Hospital Pharmacy*, pp. 75–87, Lea and Febiger, Philadelphia.

Herbst, A. L., Ulfelder, H., and Poskanzer, D., 1971, Adenocarcinoma of the vagina, association of maternal stilbesterol therapy with tumor appearance in young women, *N. Engl. J. Med.* **284**(16):878–881.

Hiranaka, P. K., and Gallelli, J. F., 1979, Investigational drugs in the hospital, in: *Handbook of Institutional Pharmacy Practice* (M. C. Smith and T. R. Brown, eds.) pp. 382–397, Williams & Wilkins, Baltimore.

Kaganov, A. L., 1980, Medical device development: Innovation versus regulation, *Ann. Thora. Surg.* **29:**331–335.

Kelsey, F. O., 1965, Problems raised for the FDA by the occurrence of thalidamide embryopathy in Germany, *Am. J. Publ. Health* **55**:703–707.

Kleinman, L. M., and Tangera, J. A., 1974, Involvement of the hospital pharmacist in single and double-blind studies, *Am. J. Hosp. Pharm.* **31**:979–981.

Kleinman, L. M., Tangera, J. A., and Gallelli, J. F., 1974, Control of investigational drugs in research hospitals, *Am. J. Hosp. Pharm.* **31**:368–371.

Lanton, R. L., 1965, A community-wide project for the collection of adverse drug reaction reports, *Am. J. Hosp. Pharm.* **22**(8):470.

Lasagna, L. 1968, General considerations concerning untoward effects of drugs in man, *Proceedings of the Third International Pharmacological Meeting* Sao Paulo, Brazil, Vol. 3 (M. Rocha e Silva, ed.), p. 84, Pergamon Press, Oxford.

Lasagna, L., and Wardell, W., 1975, The rate of new drug discovery, in: *Drug Development and Marketing* (R. B. Helms, ed.), p. 157, American Enterprise Institute for Public Policy Research.

Luy, M. L., 1976, Package Insert Roulette: The "Catch 23" of prescribing, *Mod. Med.* (Chicago) **1976**:23–20.

Mandl, F. L., Greenberg, R. B., 1976, Legal implications of preparing and dispensing drugs under conditions not in a product's official labeling, *Am. J. Hosp. Pharm.* **33**:814–816.

Stephen, S. P., and McKinley, J. D., 1974, Investigational drug information in a cancer research hospital, *Am. J. Hosp. Pharm.* **31**:372–374.

Physicians' Desk Reference, published annually by Medical Economics, Oradell, New Jersey.

9

Research Involving Medical Devices

MARY KAY RYAN

Historical Overview

Unlike the history of drug regulation within the United States, there seems to be no single, major incident to which one can attribute the genesis of medical device regulation. While there is a history of product failures and recall of devices having potentially serious consequences (Beck, 1979b), the impetus behind government regulation of medical devices seems to have been the increasing complexity of medical technology, the number of new therapeutic and diagnostic devices entering the market, and the general regulatory atmosphere prevalent in the United States. The intent has been to prevent a "thalidomide" in the world of medical devices.

There are now well over 5000 medical devices in use in the modern day hospital (Brown, 1976). Since World War II tremendous advances have been made in computers, microprocessors, prosthetic materials such as plastics or metals, solid state devices, and sensors. In 1980 the national press reported a host of experiments involving experimental medical devices: implantation of experimental insulin pumps for diabetics and miniature defibrillators to prevent attacks of ventricular fibrillation, artificial joints to replace arthritic ones, preparation of artificial heart implants, potential uses of the new positron emission tomograph (PET) scanners to detect changes in the brain's chemistry, etc. (Mannisto, 1980; Covelli, 1981). It has even been suggested that medical devices will become the "pharmaceuticals" of the next development phase of

MARY KAY RYAN ● Long Island Jewish–Hillside Medical Center, New Hyde Park, New York 11042.

medicine (Reiser, 1979) and may become preferable to drugs in managing disease if it can be demonstrated that there will be reduced incident of immediate and long-term side effects.

This increased reliance on sophisticated technology for diagnosis and treatment in modern medicine probably made regulation inevitable. In addition, many medical devices are imported from other countries where there is little or no regulation. Problems have resulted from such devices, which were not promoted as experimental, and for which there were inadequate instructions for application and/or misleading claims of efficacy.

The development of regulations by the FDA has taken a particularly arduous course. Regulations proposed on August 20, 1976 elicited widespread negative comment from industry and health professionals. The proposed regulations were characterized as having been overly restrictive and threatening to the ongoing development of new devices and advancements in biomedical science. The FDA therefore devoted considerable time and effort to the final regulations that were published in the *Federal Register* on January 18, 1980. These regulations delineate with great specificity the responsibilities of the sponsor (i.e., manufacturer of the device), the institutional review board, and the clinical investigator; they also expand upon the required elements of informed consent and specify the recordkeeping responsibilities of the researcher, IRB, and sponsor.

The Food, Drug, and Cosmetic Act amendments of 1976 set forth certain standards for medical devices regarding good manufacturing practices, labeling, performance standards, misbranding, and the use of color additives. To develop and test a new medical device, a manufacturer must be exempt from certain provisions of the 1976 Act relating to manufacturing standards, labeling, performance standards, etc. The January 18, 1980 FDA medical device regulations specify how and when sponsors of new devices may seek these exemptions and the responsibilities of clinical investigators and IRBs in the process of testing new medical devices. This chapter discusses devices that are exempt from these regulations, major procedural distinctions between research and clinical use of significant and nonsignificant risk devices, and responsibilities of the IRB, clinical investigator and the sponsor or sponsor/investigator. It also discusses some potential problems created for IRBs by these regulations.

Medical Devices Exempted from the Regulations

The following medical devices are *exempt* from the regulations:

1. Devices in commercial distribution prior to May 28, 1976, and substantially equivalent devices if used or investigated in accord with the labeling in effect at that time.
2. Diagnostic devices that comply with all applicable requirements found in the section of the *Federal Register* entitled "*In Vitro* Diagnostic

Products for Human Use" (21 CFR 809.10(c)), provided that the testing is noninvasive, that it does not require an invasive sampling procedure that presents significant risk, that it does not by design or intention introduce energy into a subject, and that it is not used as a diagnostic procedure without confirmation by another medically established diagnostic product or procedure.

3. Devices undergoing consumer preference testing, testing of a modification, or testing of a combination of devices in commercial distribution, if the testing is not to determine safety or effectiveness.

4. Devices intended solely for veterinary use or for research with laboratory animals and that are so designated in the labeling.

5. Custom devices unless used to determine safety or effectiveness for commercian distribution.

All other investigational medical devices must be reviewed by an IRB before testing is permitted.

The Role of the IRB: Determination of Risk

Before a sponsor may begin an investigation of a medical device, a duly constituted IRB must determine the degree of risk inherent in the device's use. The sponsor is directed to submit to the IRB an investigational plan including (i) the name and intended use of the device; (ii) the objectives and duration of the investigation; (iii) a written protocol describing the methodology to be used and an analysis of the protocol demonstrating that the investigation is scientifically sound; (iv) a description and analysis of all *increased risks* to which subjects will be exposed; (v) the manner in which risks will be minimized; (iv) a justification for the investigation; (vii) a description of the patient population including information on the number of subjects to be recruited, their age, sex, and condition; (viii) a description of the device that should include each important component, ingredient, property, principle of operation, and anticipated changes in the device during the course of the investigation; (ix) a written procedure for monitoring the investigation; (x) a copy of all labeling to be used for the device; (xi) consent forms and informational material to be presented to the subjects; (xii) a list of all IRBs which have been or will be asked to review the investigation, including names of the chairpersons and locations, and certification of any action taken by any IRB regarding the investigation; (xiii) the name and address of each institution at which a part of the investigation may be conducted; and (xiv) a description of records and reports which will be maintained on the investigation in addition to consent forms and informational materials given to subjects.

The sponsor must also provide the IRB with (i) a report on prior clinical, animal, and laboratory testing of the device which should be comprehensive and adequate to justify the proposed investigation; (ii) a bibliography of adverse and supportive publications relevant to an evaluation of the safety or effectiveness of the device, copies of all published and unpublished adverse information, and copies of significant publications if requested by an IRB or FDA; (iii) a summary of all other unpublished information (whether adverse or supportive) relevant to an evaluation of the safety or effectiveness of the device; (iv) a statement that all nonclinical tests have been conducted in compliance with applicable requirements in the good laboratory practice regulations.

Based on this extensive information, the IRB is expected to evaluate the scientific soundness of the project, the risk/benefit ratio, and ethical considerations relevant to patient recruitment and to make a determination as to whether the device presents a nonsignificant or significant risk to subjects. The IRBs should note that there is currently no provision in the regulation for an expedited review of nonsignificant risk devices.

Nonsignificant Risk Devices

If the IRB determines that the device does not pose a serious risk to subjects, the sponsor/investigator may immediately begin to test or use the device in therapeutic trials. Nonsignificant risk devices include crutches, elastic knee braces, bedboards, bedpans, medical chairs, and tongue depressors as ordinarily used. Specimen containers, e.g., blood or sputum collection devices, would also appear to fit into this category. Unless otherwise informed by the FDA, a sponsor or sponsor/investigator is approved to conduct investigations on nonsignificant risk devices if it is not a banned device and if the sponsor (i) labels the device in accordance with FDA regulations; (ii) has IRB approval after presenting the IRB with the reasons why the device poses no serious risk; (iii) maintains IRB approval throughout the investigation; (iv) ensures that investigators obtain and document informed consent for each subject; (v) complies with FDA regulations regarding monitoring of the investigations, maintaining records, and filing reports; (vi) ensures that participating investigators maintain all required records and file required reports; and (vii) comply with prohibition on promotion, test marketing, and commercialization of the investigational device. The FDA has stated in the preamble to the regulations that "the IRB serve as the surrogate of the Secretary with respect to the receipt and approval of the application." If an IRB disagrees with the sponsor's categorization of the device as one of nonsignificant risk, the sponsors are required to notify the FDA and submit an application to the FDA to investigate the device as one positing significant risk to subjects.

Significant Risk Devices

Federal regulations define a significant risk device as one that "presents a potential for serious risk to the health, safety, or welfare of a subject." These are devices used to support or sustain human life, devices considered to be of substantial importance in diagnosis and treatment of disease and in preventing impairment of human health, and implants, which are defined as devices "placed into a surgically or naturally formed cavity of the human body if intended to remain there for a period of thirty days or more." The FDA has reserved the right to categorize specific implants placed in subjects for shorter periods as significant risk devices.

The process whereby the sponsor applies to the FDA for permission to carry out an investigation of a significant risk device involves the IRB at the outset. *Before* submitting an application to the FDA for approval to conduct such an investigation, the sponsor or sponsor/investigator must first obtain IRB approval for the study. If no IRB exists or if the FDA finds the IRB review inadequate, the sponsor may apply directly to the FDA for approval to conduct the research. (The FDA, however, may refuse to approve the sponsor's application without IRB prior approval if it feels that IRB review is necessary for the protection of human subjects.) Sponsors and investigators are prohibited from beginning an investigation involving a significant risk device until both the local IRB and the FDA have approved it. The FDA will notify the sponsor or sponsor/investigator of the date it receives the application. *The investigation may begin thirty days after the FDA receives the application unless the FDA notifies the sponsor to the contrary.*

It is likely that a sponsor will test a new device in more than one institution. In this case, the sponsor is required to submit to the FDA updated information on each IRB reviewing the application, certification of IRB approval, and a description of any modifications in the investigational plan required by an IRB as a condition of its approval.

Responsibilities of the IRB, Sponsor, and Clinical Investigator

The FDA has delineated the responsibilities of IRB, clinical investigators, and sponsor in the medical device regulations to a degree not presently in evidence in regulations governing other areas of biomedical research (including drugs).

Responsibilities of the IRB

The medical device regulations call for an IRB constituted essentially in the same manner as the regulations set forth by the Department of Health and Human Services (HHS). The IRB must consist of at least five members of diverse backgrounds and must include a licensed physician, a nonphysician scientist, a representative of a nonscientific field, and a member who is not affiliated with the institution. The IRB is charged with reviewing the proposal for an acceptable risk/benefit ratio and for acceptability in terms of community attitudes, ethical standards, and professional conduct [HHS regulations, section 812.62(a)–(d)].

There are, however, several added imperatives. The IRB "shall have among its members or shall obtain by means of nonvoting consultants sufficient scientific and technical knowledge and expertise to be able to review proposed investigations . . ." [HHS regulations, section 812.62 (e).] Further, "no IRB, institution, or other person may permit an investigator or sponsor to participate in the selection of members of an IRB that will review an investigation conducted or sponsored by that investigator or sponsor." [HHS regulations, section 812.62 (8).] These two requirements will place burdens on many smaller community hospitals.

These regulations further require that the IRB have a written procedure for conducting its review and for reporting its decision to the institution, investigator, and sponsor of the device. The review must be conducted by a quorum, and this quorom must consist of a licensed physician, a nonphysician scientist, and one member representing a nonscientific area. The IRB must ensure against conflict of interest, must make decisions in a timely manner, and must notify the investigator (and, where appropriate, the sponsor) of each decision it makes about the investigation and the basis for its decision.

The regulations also specify that the IRB must maintain the following records: (i) detailed minutes of each meeting and of each investigation; (ii) all correspondence with another IRB, an investigator, a sponsor, a monitor, or the FDA; (iii) records on IRB membership including relationship to the institution, earned degrees (if any), occupation, etc; and (iv) minutes of attendance at each meeting. These records must be maintained during the investigation and for two years after the latter of two dates: the date on which the investigation is terminated or completed, or the date that such records are no longer required for purposes of supporting a premarket approval application or a notice of completion of a product development protocol.

Responsibilities of the Investigator

To an extent not presently existing in other regulations, the medical device regulations impose a number of obligations and responsibilities on investigators participating in clinical trials. The investigator is responsible for conducting the trial in accordance with the signed agreement, the investigational plan, and FDA regulations, and for protecting the rights and welfare of patients enrolled in the study. The investigator must obtain informed consent, is responsible for the control of the device, may not supply a device to a person other than those authorized to receive it, and must return devices to the sponsor upon completion or termination of the investigation.

These regulations also impose very specific and, unfortunately, duplicative record-keeping requirements (Table I). The investigator must retain (i) all correspondence with the IRB, with the sponsor, and with the FDA, including reports (the IRB and sponsor are called upon to maintain identical files); (ii) records on receipt of the device including code mark or batch numbers and

Table I. Responsibilities for Maintaining Records[a]

Records	Maintained by		
	Investigator	Sponsor	IRB
Significant risk device			
All correspondence pertaining to the investigation	✓	✓	✓
Shipment, receipt, disposition	✓	✓	—
Device administration and use	✓	—	—
Subject case histories	✓	—	—
Informed consent	✓	—	—
Protocols and reason for deviations from protocol	✓	—	—
Adverse device effects and complaints	✓	✓	—
Signed investigator agreements	—	✓	—
Membership/employment/conflicts of interest	—	—	✓
Minutes of meetings	—	—	✓
Nonsignificant risk device			
Name and intended use of device	—	✓	—
Brief explanation of why device does not involve significant risk	—	✓	—
Name and addresses of investigator(s) and IRBs	—	✓	—
Degree GMPs followed	—	✓	—
Informed consent	✓	—	—
Adverse device effects and complaints	—	✓	—

[a]From An Overview of the Investigational Device Exemption Regulation, DHHS Publication No. (FDA) 80-4023.

receipt, date, names of persons who have used or disposed of each device; (iii) information on why and how many units of the device were returned or repaired or otherwise disposed of; and (iv) records of each subject's case history and exposure to the device including informed consent, or justification for using the device without obtaining informed consent. The investigator's files (which are subject to FDA audit and to monitoring by sponsor and IRB) must also contain all relevant observations, including records concerning adverse effects (whether anticipated or unanticipated), information and data on the condition of each patient upon entering and during the course of the investigation, as well as information about the patient's relevant previous medical history and the results of all diagnostic tests. Records must be kept by the investigators concerning the exposure of each subject to the device, the time and date of each use, and any other therapy. The protocol must be kept on file with documents showing the dates of and reasons for each deviation from the protocol.

Table II. *Responsibilities for Preparing and Submitting Reports for Nonsignificant Risk Devices*[a]

	Report prepared by	
Type of report	Investigators for	Sponsors for
Unanticipated adverse effect evaluation	Sponsors and IRBs	FDA, investigators and IRBs
Withdrawal of IRB approval	Sponsors	FDA, investigators and IRBs
Progress report	N/A	FDA[b] and IRBs
Final report	N/A	IRBs
Inability to obtain informed consent	Sponsors and IRBs	FDA
Withdrawal of FDA approval	N/A	IRBs and Investigators
Recall and device disposition	N/A	FDA and IRBs
Significant risk determinations	N/A	FDA

[a] From *An Overview of the Investigational Device Exemption Regulation*, DHHS Publication No. (FDA)80-4023.
[b] FDA is planning to drop the reporting requirement for submitting progress reports to FDA for nonsignificant risk device investigations.

The investigator must also keep any other records that the FDA may require or that are required by virtue of a particular clinical trial.

Informed Consent Requirements

The investigator must obtain informed consent from prospective subjects. There is no provision that anyone other than the investigator (i.e., a clinical or research fellow) may be delegated to obtain informed consent from the patient. In addition to the detailed record-keeping requirements, these regulations specify expanded elements of information that the investigator must provide the patient during the informed consent procedure. The investigator must inform the patient that the device is being used for research purposes. The prospective subject must be given an explanation of the likely results should the procedures fail, and must be told the scope of the investigation and the number of subjects involved. Each of these three expanded elements of informed consent may pose particular difficulties for the investigator.

There are also three notable omissions from the informed consent requirements of these regulations. In spite of the expansion in the elements of informed consent, the regulations on medical devices omit the requirement that prospective subjects be informed of what, if any, compensation is available to them should they suffer an injury as a result of their participation in the research. This is in marked contrast to the HHS regulations. With all the record-keeping requirements imposed on the investigator relating to reporting unanticipated adverse effects, there is no requirement that the investigator pass along to subjects information that may affect the subject's decision to continue

Table III. Responsibilities for Preparing and Submitting Reports for Significant Risk Devices[a]

Type of report	Report prepared by	
	Investigators for	Sponsors for
Unanticipated adverse effect evaluation	Sponsors and IRBs	FDA, investigators and IRBs
Withdrawal of IRB approval	Sponsors	FDA, investigators and IRBs
Progress report	Sponsors, monitors and IRBs	FDA and IRBs
Final report	Sponsors and IRBs	FDA, investigators and IRBs
Emergencies (protocol deviations)	Sponsors and IRBs	FDA
Inability to obtain informed consent	Sponsors and IRBs	FDA
Withdrawal of FDA approval	N/A	IRBs and investigators
Current investigator list	N/A	FDA
Recall and device disposition	N/A	FDA and IRBs
Records maintenance transfer	FDA	FDA
Significant risk determinations	N/A	FDA

[a] From An Overview of the Investigational Device Exemption Regulation, DHHS Publication No. (FDA) 80-4023.

participating. In further contrast to HHS regulations, there is no requirement that the prospective subject be informed to what degree confidentiality of records can be maintained. These are, however, areas of concern to IRBs and must be included in the IRB's review and directive to investigators.

Responsibilities of the Sponsor and Sponsor/Investigator

The sponsor or manufacturer's responsibilities would be of less interest to those concerned with IRB review of medical devices were it not for the fact that the sponsor (manufacturer) could also be the physician/investigator. Practicing physicians and dentists have been active both in the development of new devices and in making significant modifications to existing technology. In such cases, the sponsor/investigator is responsible for meeting the FDA's requirements for both sponsors and investigators relative to submitting a protocol in conformity with the regulations for IRB review, for record-keeping, and the conduct of the project. Reporting requirements for investigators and sponsors for nonsignificant and significant risk devices are detailed in Tables II and III. The timing of these reports are also mandated by these regulations (see Tables IV and V).

Areas of Concern

The problems facing IRBs and investigators as a result of these regulations fall into several categories: informed consent; confidentiality of medical

Table IV. Timing of Reports for Significant Risk Devices

Type of report	Report from sponsor for	Time limit	Report from Investigator for	Time limit
Unanticipated adverse device effects	FDA, IRBs, investigators	10 working days of first notice	Sponsor, IRB	10 working days of first learning of it
Withdrawal of IRB approval—all or in part	FDA, IRBs, investigators	5 working days of receipt	Sponsor	5 working days of receipt
Withdrawal of FDA approval	IRBs, investigators	5 working days of receipt	NA	—
Progress reports	FDA, IRBs	Regular intervals, at least yearly	Sponsor, monitor, IRB	Regular intervals, at least yearly
Current investigator list	FDA	6 month intervals, beginning 6 months after commencement of study	NA	—
Recall and disposition of devices if sponsor request return of devices	FDA, IRBs	30 working days of request, including reasons for the request	NA	—
Use of device without informed consent	FDA	5 working days of receipt of information	Sponsor, IRB	5 working days of use
Deviations from investigational plan	NA	—	Sponsor, IRB	5 working days of emergency (prior approval required if no emergency)
Significant risk device determinations	FDA	5 working days of learning of IRB determination	NA	—
Final report	FDA, IRBs, investigators	30 working days of completion, termination	Sponsor, IRB	3 months of termination

Table V. Timing of Reports for Nonsignificant Risk Devices

Type of report	Report from sponsor for	Time limit	Report from investigators for	Time limit
Unanticipated adverse effects	FDA, IRBs, investigators	10 working days of first learning	Sponsor, IRB	10 working days of first learning
Withdrawal of IRB approval—all or in part	FDA, IRBs, investigators	5 working days of receipt	Sponsor	5 working days of notice
Withdrawal of FDA approval	IRBs, investigators	5 working days of receipt	NA	—
Progress reports	IRBs	Regular intervals, at least yearly	NA	—
Recall and disposition of recalled devices	FDA, IRBs	30 working days of request, including reasons for request	NA	—
Use of device without informed consent	FDA	5 working days of receipt	Sponsor, IRB	5 working days of use
IRB determination that device is a significant risk device	FDA	5 working days of first learning	NA	—
Final report	IRBs	6 months of termination	NA	—

records, procedural difficulties, the ability of IRBs to discharge the responsibility imposed by the FDA, problems inherent in the nature of certain devices such as how and by whom they will be actually used, and the possible impact on future developments in medical technology.

Informed Consent

The HHS regulations governing IRBs and the elements of informed consent mandate that subjects be informed what, if any, compensation is available to them for physical injuries resulting from participation in research. This requirement is conspicuously absent in the FDA regulations on devices. Local IRBs should probably direct investigators to include this information in the informed consent process. It would appear that if such information is considered relevant to all other research, it cannot be deemed irrelevant to medical device investigations.

Holder (1980) cites several problems with the directives to expand the elements of informed consent. The investigator is directed to inform the subject "that the investigational device is being used for research purposes." More often than not the device will be used as well for therapeutic or diagnostic purposes. The investigator is directed to inform the subject of the likely results should the procedure fail. It can be argued that in cases where the subject is critically ill, this information could be harmful. The subject must also be informed of the scope of the study, including the number of subjects to be enrolled. This information is irrelevant; it is more important to know whether one is the fifth or twenty-fifth in a study of thirty subjects and the incidence of complications. Furthermore, with devices used in critical cases, it may not be possible for the physician to predict how many times he can expect to use the device.

Confidentiality

The FDA regulations pose serious difficulties for IRBs and investigators who are required by HHS regulations to maintain confidentiality of patient records (Hoffman, 1979; Clark, 1979; Gildenberg, 1980). FDA investigators are authorized to inspect IRB records and *to copy all records relating to the investigation*. The investigator must permit the FDA access to inspect and copy records that identify subjects if the FDA "has reason to suspect that adequate informed consent was not obtained" or if other aspects of the investigation are in doubt. These regulations may result in the loss of proper security for health records; details of patient charts would become part of official records and the Freedom of Information Act would permit disclosure on request. Hoffman (1979) warned "there may not be any evaluation of medical devices related to management of patients with veneral disease, drug abuse, or devices related to any disease which an individual may perceive as injurious to chances for employment and promotion, insurance, or immunity from legal action." It has also been pointed out that under FDA regulations IRBs may be faced with problems of confidentiality of information about a new device when patents and proprietary rights are involved (Dobelle, 1977).

Procedural Difficulties

There are several unnecessary procedural difficulties created by these regulations. An investigator and his patient will face a delay of thirty days following IRB approval while awaiting FDA action before being permitted to use a device classified as significant risk. Since such devices will most probably be used on seriously ill patients, this delay is of concern to many physicians (Gildenberg, 1980). Since the IRB will have already judged the device as one of potential benefit, and will have approved a consent document, this procedural delay is difficult to justify. In emergencies and in cases where there is no medically accepted alternative, the device may be used immediately. There are too many cases, however, which cannot be classified as emergencies and for which there may be an alternate therapy. Since a significant risk device is most likely to be used in serious illness where the potential for litigation exists in cases of misuse, any institution would likely be hesitant about liberally conferring the status of emergency on cases to avoid the thirty day wait period required by law.

Another procedural difficulty is the absence of a provision for expedited review for insignificant risk devices. In this case, one would hope that common sense will prevail. There are a host of insignificant risk devices that could be reviewed effectively by the IRB through an abbreviated process: new tongue depressors, crutches, adhesive bandages, knee braces, wheelchairs, etc. Since the final HHS and FDA regulations provide for expedited review, an IRB may decide to implement its procedure for expedited review in the same context as it is applied to other clinical investigations.

The FDA is apparently concerned about conflicts of interest occurring among the IRB, sponsor, and clinical investigator and IRB review of medical devices. The regulations state that "no IRB, institution, or other person may permit an investigator or sponsor to participate in the selection of members of an IRB that will review an investigation conducted or sponsored by that investigator or sponsor." This rule displays an ignorance about the ways in which institutions select IRB members. It is highly likely that chairmen of departments and/or chiefs of divisions in medicine, surgery, orthopedics, radiology, laboratories, etc., will appear as the investigator or coinvestigator on a protocol requesting approval to use a new medical device in research and clinical care in their department. These are the same individuals who are routinely asked to nominate and/or approve appointments of their faculty for IRB membership.

It will simply be impossible for most institutions to comply with regulations prohibiting investigators from participation in IRB selection. The regulations further require that the investigator obtain informed consent from prospective subjects; there is no provision that any other physician member of the research or clinical team may obtain informed consent. One suspects that those who wrote the regulations are unfamiliar with the conduct of clinical medicine and research in hospitals where teams of health professionals work together on projects and procedures. There will be many cases where the investigator of record will be unavailable to obtain informed consent while other physicians and clinical and research fellows would be able to do so.

The Ability of the IRB to Review Medical Devices

Legitimate questions about the ability of the average IRB to discharge its responsibilities under these regulations can be raised.

The protocol that the IRB may review will contain very technical information including "a description of the device which should include each important component, ingredient, property, principle of operation, and anticipated changes in the device during the course of the investigation." The IRB is being called upon to assess not only the basic scientific validity of the research project but also the risks inherent in the device itself. Few institutions have access to the array of engineering specialties or professionals with expertise necessary to make such an assessment. Yet the regulations direct IRBs to have among its members, or to obtain via consultants, individuals with sufficient scientific technical knowledge and expertise to be able to review these protocols. Major academic medical centers and medical and dental schools which are part of large universities may have no problem in reviewing the technical information presented by the sponsors or manufacturer of the device. Smaller institutions and community teaching hospitals may have to resort to hiring consultants in order to meet the requirements of the regulations. When one considers that some sponsors will need to test a significant risk device at several hospitals, each of which may have hired a consultant to perform the identical evaluation, one can see that the whole process could become rather costly for the health care system.

The Nature of Clinical Devices

Unlike drug research, where protocols can specify dosage and attempt to forecast reactions and side effects, the successful use of a particular medical device may depend on the skill of the individual who inserts, implants, or applies it or who interprets the data generated by a new diagnostic device. In addition, some devices will be used by the patient at home in situations where direct professional supervision and consultation will not be available. This should be considered by the IRB insofar as it must review the instructions for use that accompany the device so as to ensure that they are intelligible to the user. Considering the variety of backgrounds, the reading skills, and the level of education of the general patient population the IRB should alert researchers to situations posing increased risk to patients responsible for the appropriate use of a new device.

Impact on Future Developments in Medical Technology

Critics of the medical device regulations have pointed to the added costs generated by the requirements for protocols, review by multiple IRBs, and increased paperwork. Clark (1980) estimated additional costs of $50,000 for IRB conformity with these regulations. The costs to manufacturers for clinical studies, the time requirements, and the potential problems have been analyzed by Frisch (1980) who maintains that continued development of moderate to significant risk devices will be hampered by these regulations. At best, devices will continue to be developed but at a greater cost—a cost eventually borne by the patient. Clark (1979) summarizes the fears of his colleagues in the following paragraph:

> There will be fewer devices in the future because of markedly increased costs of development and bureaucratic approval. Devices currently available and all future devices will be more expensive to patients, insurance carriers, and government. The availability of special order or custom-made devices for specific patients will decrease. The act discourages clinical investigation because of the increased costs, time required, and risk imposed on the investigational surgical practitioner. Surgeons will be required to notify all patients concerning a potential problem with a device, which may cause unnecessary anxiety. The surgical practitioner in small institutions will not have use of investigational devices because of the restrictions imposed by the proposed regulations and the authority of the Secretary of the FDA to limit the use of investigational devices to specific centers and surgeons. The innovative small manufacturer who has produced so many of the new technology devices will generally disappear from the marketplace because of an inability to meet the regulatory restrictions due to limited capital and resources. The proposed regulations will put voluntary standards and certification organizations under federal control. There will be a large increase in time and costs to institutions for the activities of institutional review boards. Importantly, these regulations will not improve the scientific data base nor significantly decrease the risk of new devices. These regulations will put more lawyers to work at the expense of the clinician, his patients, and the American public.

Conclusion

The impact of these regulations on the development of new technology is not the primary concern of IRBs. In defense of the FDA, the regulations streamline the review process for testing and therapeutic use of nonsignificant risk devices. While the regulations appear overly burdensome with regard to devices of significant risk, much of the burden may be alleviated by the investigator and manufacturer making themselves available to resolve the IRB's questions.

Industry has been generally resistant to regulation and, in some cases, with good cause. There have been, however, serious problems caused by failures of investigational devices. While product failures in other industries may be inconvenient to the consumer, when the product is a new pacemaker, a new intraocular lens, an implanted insulin pump, or a new diagnostic device failure can cause fatalities or permanent disabilities.

The unease with which IRBs received these regulations is based on a few major concerns. First, IRBs do not regard themselves as surrogates for the federal government. Second, many IRBs feel inadequate to review medical devices. A fully developed protocol for a significant risk device will contain information that only a highly qualified engineer familiar with the technology employed by the device could evaluate. Such individuals are not readily available to IRBs. Furthermore, recent events have made IRBs cautious about nonsignificant risk devices. At the time these regulations were going into effect, a major pharmaceutical company withdrew a new product, a tampon, from the market because it became associated with the death of thirty women who used it. Had this product been submitted to an IRB for review, it probably would have been categorized as one of nonsignificant risk. These events shake the confidence of many IRBs and lead to an overly cautious stance. Finally, confidentiality of patient records is a fundamental concern of physicians and hospital administrators. These regulations authorize FDA inspectors to copy all records relating to the research if it has doubts about informed consent or the data. For many IRBs, physicians, and administrators this is too sweeping an authority.

It is probably too early to judge how these regulations will effect medical device research and the continued development of medical knowledge. Revisions and modifications in federal regulations are not uncommon. Whether or not improvements will be made in these regulations will depend a great deal on the feedback that the FDA receives from IRBs, sponsors, and physicians, both as individuals and through professional organizations (Beck, 1979a).

References

Beck, W. C., 1979a, The physician's role in the quality control of medical devices, *J. Am. Med. Assoc.* **241**:56–57.

Beck, W. C., 1979b, The surgeon's role in device standardization, *Am. J. Surg.* **137**:149–151.

Brown, J. H. U., 1976, Medical devices as medical problems: The regulation of devices for clinical use, *Biomed. Eng.* **11**:337–339.

Clark, R. E., 1980, Medical device regulation: Current and future trends, *Ann. Thorac. Surg.*, **29**:298–299.

Clark, R. E., 1979, Federal regulation and the practicing surgeon, *Am. J. Surg.* **138**: 775–778.

Covelli, P., 1981, New hope for diabetics, *The New York Times Magazine* **1981** (March 8):62–70.

Dobelle, W. H., 1977, Minimizing the adverse effects of the medical device amendments of 1976 on innovation in artificial organs research, *Artific. Organs* **1**:65–75.

Frisch, E., 1980, An economic evaluation of premarket regulatory requirements, *Med. Devices Diagnostic Industry* **2**:31–38.

Gildenberg, P., 1980, Neurosurgical devices and drugs, *Neurosurgery* **6**:220–223.

Hoffman, F., 1979, Has the FDA provided the best possible solution? *Man Med.* **4**:198–200.

Holder, A., 1980, FDA's final regulations: IRBs and medical devices, *IRB: A Review of Human Subjects Research* **2** (6):1–4.

Mannisto, M., 1980, Public press focuses on advances in medical research/technology, *Hospitals* **1980** (December 16):65–68.

Reiser, S. J., 1979, Striking a balance, *Man Med.* **4**:201–202.

Vladeck, B. C., 1979, Regulation and innovation, *Man Med.* **4**:203–204.

10

Continuing Review of Research

STUART WOLLMAN AND MARY KAY RYAN

Overview

The government has been slow to address itself to the question of monitoring ongoing research. In 1966 the Surgeon General acting on the Kefauver–Harris amendments of 1962, requested, and most research publications in turn required, investigators' assurances that informed consent had been obtained. In the 1971 Institutional Guide to HEW policy, IRBs were charged with establishing a basis for continuing review in keeping with initial review determinations of risk/benefit, rights and welfare of subjects and informed consent. In 1974 HEW (*Federal Register,* section 46.2(4), Volume 39, No. 105, p. 18917) stipulated that where the board "finds risk is involved . . . , it shall review the conduct of the activity at timely intervals."

The Food and Drug Administration (FDA) had its own set of regulations which, while not identical to HEW's, also called for review to be timed on the basis of the degree of risk but not less than once a year. The FDA also gave the IRB the authority to suspend or terminate a research project, which HEW regulations did not.

Although HEW required institutional review board approval prior to the initiation of clinical investigations as early as 1974, it was not until mid-1977 that the FDA conducted a pilot program that inspected institutional review committees. This pilot program explored procedures for continuing review, frequency of reviews, personnel involved, actions taken, and the length of time records were retained. Investigators' reports of unexpected side effects, alarm-

STUART WOLLMAN ● Long Island Jewish–Hillside Medical Center, New Hyde Park, New York 11042, and Department of Clinical Anesthesiology, State University of New York at Stony Brook, Stony Brook, New York 11794. MARY KAY RYAN ● Long Island Jewish–Hillside Medical Center, New Hyde Park, New York 11042.

ing adverse reactions, subject deaths, and the IRB response to these reports were all studied. Until this point, IRBs judged, according to their own standards, what the review would consist of, who would review projects, how it would be carried out, and what the 1974 HEW phraseology regarding "conduct of the activity" and "timely intervals" meant.

The first sign that this latitude would be curtailed appeared in 1978, when the National Commission on the Protection of Human Subjects of Biomedical and Behavioral Research issued a recommendation calling for closer monitoring of the conduct of approved research and monitoring of the consent process. The National Commission was established in July, 1974 following Senate hearings conducted by Senator Edward M. Kennedy during which several controversies over approved research projects came to light, including sterilization without informed consent, NIH-sponsored metabolic research using decapitated fetal heads, and the Tuskegee Syphilis Study in which 300 males participated in a syphilis treatment research project (most of them black and of lower educational levels), without their informed consent, long after effective treatment had become available.

The National Commission concluded its work in October 1978. During its tenure ten reports were issued along with several recommendations for improved IRB functioning. Recommendations 3(D) and (E) called for IRBs to "have the authority to conduct continuing review of research involving human subjects and to suspend approval of research that is not being conducted in accordance with the determination of the Board or in which there is unexpected serious harm to subjects;" and to "maintain appropriate records, including copies of proposals reviewed, approved consent forms, minutes of Board meetings, progress reports submitted by investigators, reports of injuries to subjects, and records of continuing review activities." The federal government heeded these recommendations and in January 1981 effected regulations that detailed the continuing review and consent process.

Controversy over Review Requirements

The final rules have increased the tension between the academic community and government, and in some cases between the research scientist and the IRB. Many feel that these regulations unnecessarily extend federal bureaucracy into research efforts on university campuses and medical centers in ways that will not automatically provide better protection for participants in research.

There is concern that the government is seeking to turn the institutional review board into an extension of its punitive authority and is failing to take into account the environment in which academic and university research takes place. Robert J. Levine, M.D., Chairman of Yale University's IRB, and consultant to the former National Commission for the Protection of Human Subjects, stated:

> We must resist any efforts to turn the IRB into a police force. Current monitoring activities can only indicate to the community of investigators that they are operating from a presumption of mistrust ... "Policing" erodes one of the basic assumptions that forms the foundation of life within a university community—that persons are to be trusted until contrary evidence is brought forward. Further, if the IRB is perceived as a police force, it will lose what I refer to as its "informal monitoring system," that is, reports brought forward by students, nurses, physicians and so on. (Levine, 1979.)

FDA and HHS regulations direct the board to inform the researcher, the institutional officials, and the government of the reasons for suspension or termination of funded research and serious violations of IRB determinations by a researcher. The board is given the authority to observe the consent process or appoint a third party to observe how consent is obtained. FDA and HHS received significant adverse comment on the continuing review aspect of the regulations proposed in 1979. Both agencies felt it necessary to discuss these comments and their position in the preamble to the final regulations announced in January, 1981 (*Federal Register,* January 26 and 27, 1981). While these rules were under discussion in the fall of 1979 and throughout 1980, the press and scientific journals carried reports of fraud in connection with research and conflicts between researchers and their IRB. Previously, such reports emanating from respected academic centers were so rare as to be almost unheard of. IRBs and institutions are now giving some thought to investigative procedures and the question at what point and with what degree of institutional concurrence the IRB should notify the government of serious violations of research ethics.

In response to the criticism that these regulations force IRBs into a "police role," the FDA responded (*Federal Register,* January 27, 1981, p. 8967) that the agency "considers it an appropriate requirement that IRBs develop procedures to determine whether there is need for verification, from sources other than the investigator, that there has been no material change in certain protocols since the previous review." Independent verification is, however, not a requirement but the FDA feels that IRBs should be aware that it is available to them in conducting their review. Furthermore, when an investigator's noncompliance is serious enough, the FDA points out to the board that "disciplinary action against the investigator may also be in order. Consequently, FDA has required ... that the IRB report an investigator's serious noncompliance to the bodies that have the authority to take action against the investigator— the institution and FDA." (*Federal Register,* January 27, 1981, p. 8967.) HHS has reiterated this position in its final regulations explaining that "the reporting to (HHS) of the suspension or termination of research is important since HHS has an obligation to examine problems associated with research supported by public funds. . . ." (*Federal Register,* January 26, 1981, p. 8377.)

These proposed regulations add a dimension to the IRB that has made board members, institutional officers, and researchers uncomfortable. An IRB in a routine continuing review procedure (such as reviewing informed consents, discussing progress of the research, and reviewing side effects) is unlikely to uncover serious violations of research ethics. Most recently, such violations have been brought to the attention of institutional officials by colleagues, students, nurses, patients, etc. Many feel that it is not the responsibility of the board to report *directly to the government* its decisions to terminate or suspend a study as a result of violations. Such charges carry with them potential liability for the accused, accusor, the institution, and IRB, and many IRBs object to being required to trigger institutional review by a federal agency. IRBs are not entirely independent of the institution that has appointed their membership and determines their tenure. Members are usually made up of faculty, administrative staff, students, and, in the case of hospitals, patient and community representatives from a variety of backgrounds. Generally speaking, faculty and administrative staff are in the majority. Up to now the procedure for dealing with a recalcitrant investigator, who may be unwilling to follow the institutions's policy governing research involving human subjects, has been to inform the appropriate institutional official. This is usually the provost, vice president for research, dean of faculty, or the president, any one of whom have the responsibility and authority to enforce institutional policy. This has been enough to end the violations.

While academic life may be characterized by scientific rivalry, ego drive, and personal ambition for advancement and promotion, the academic process is also characterized by open disclosure and honest evaluation of results. Anyone familiar with academic institutions can attest to the degree of freedom any faculty member has in publicly questioning and challenging a colleague's work. Some of the worst abuses in human experimentation have taken place in situations that were far removed from the academic environment and where there has been no real opportunity for interdisciplinary discussion.

Many fear that the regulations will make the IRB an enforcement arm of the FDA, which already has the authority to audit any institution falling under its jurisdiction. Until now "continuing review" might have consisted of a questionnaire to the researcher requesting certification that the project was still underway, that the approved consent form was still in use, and that no untoward effects had been experienced by participants in the study. The tenor of the current regulations, which give the board the right to appoint a third party observer to judge the quality of the consent process and/or the conduct of the investigator, could bring on a mood of confrontation and distrust that will not necessarily protect subjects any better than an open dialogue between investigator and the board (Levine, 1979). IRBs should use this right sparingly.

Benefits of the Review Process

The reports and recommendations of the National Commission and subsequent proposals by HEW and FDA attempt to set an objective standard against which protocols should be evaluated. Rules regarding membership, the definition of a quorum, and the elements the board must consider in evaluating the consent process have all become more specific. The net effect is that IRB discretion has become limited. Most would agree, however, that the continued assessment of research involving human subjects is essential to the protection of human subjects and to the maintenance of quality in research. Review boards must now grapple, if they have not already done so, with conducting continuing review in ways that will insure protection of human subjects and, hopefully, the furtherence of ethical research.

In biomedical research monitoring untoward events, morbidity, and mortality on a regular basis may aid in the early detection of a trend that could mandate a change in or termination of a study. The presence of a fresh point of view, personified by selected reviewers, could be useful and valuable assistance to an investigator whose close involvement in the original design of the study may make him less sensitive or objective to subtle trends. If the review team is appropriately constituted, the federally mandated continuing review process becomes a means to further improve academic research through exposing investigators to the opinions and suggestions of their colleagues on a regular basis.

The process of continuing review primarily serves the important purpose of protecting those who participate in research. Review boards are usually intent on reviewing many research projects, on matching up procedures described in the proposal with the explanation appearing in the consent form, and in assessing what, if any, risks participants face. They often fail to assess exactly what their colleagues may think about the role they play in research and what their colleagues may fundamentally understand about the requirements for human experimentation.

Generally speaking, most researchers are aware of the need for a consent form and for prior review and approval of projects. Many scholars and scientists, however, have never been exposed to these present day requirements in any organized way. Few doctorate programs or medical or dental schools include the history of human experimentation in their curriculum. The process of continuing review can alert the review board to both individual and collective gaps of knowledge in their institution. Institutional seminars, forums, etc., can educate the scientific community about this important area and these, in the long run, will improve compliance, increase sensitivity, and provide further safeguards to protect human subjects in research.

Procedures for Reviewing Research

Continued assessment of research involving human subjects is essential. The consideration of a protocol for its risks, its benefits, its candor, and clarity of its consent form is only the first step in the protection of human subjects. It is the responsibility of an IRB to be aware of investigator compliance with established procedure changes in experimental protocols, untoward reactions, morbidity, mortality, and preliminary results that may alter earlier risk/benefit assumptions. Heath (1979) has discussed the various aspects of monitoring human research. Heath (1979) proposed that the monitoring of research on human subjects address such considerations as continuing review, consent review, adherence to protocol, and identification of unapproved activities. Audit procedures must address these concerns to properly safeguard subjects.

The need for a review mechanism has been established by the studies of Gray (1975) who demonstrated noncompliance with or evasion of review of committee directives in a national sample of medical institutions. In this study of 66 hospitals, Gray found that 9% of the investigators never submitted their projects for institutional review and approval. Of those who obtained board approval, 13% failed to use their institution's required consent forms and did not even inform subjects that they were involved in research. Of those subjects who signed proper consent forms, 30% denied knowledge of involvement in research when interviewed 1–2 days after signing a consent form.

These findings make it clear that while it is important to review compliance by evidence of signed consent forms, reviewers are obviously faced with the problem of assessing the spirit of compliance with institutional review board requirements to inform and involve subjects in the studies for which they have been recruited.

Most investigators' priorities are the scientific questions addressed in their projects. Beecher (1966) showed numerous instances where these concerns have blinded investigators to the risks experimental subjects were facing. Compliance with institutional review board directives is regarded by many as a nuisance or a source of delay. Many investigators, totally unaware of the events that led to federal regulations governing research and seeing themselves as doing valuable work for the good of mankind, are offended when someone looks over their shoulders to observe how subjects are treated. Society has judged, however, that those who participate in experiments need the protection of a disinterested, informed, and sensitive institution to review human research.

The periodic review of ongoing research is an opportunity for assessment of preliminary data which may change risk/benefit assessments made at the start of the study. The monitoring of untoward events, morbidity, and mortality on a regular basis may aid in the early detection of a trend that could mandate a change in or termination of the study. The presence of a fresh point of view, personified in the reviewers, may be a useful and valuable assistance to an investigator whose close involvement in the original design of his study may make him less sensitive to subtle trends. The proper review of ongoing research can help minimize abuse and offers benefits to investigators as well as to subjects.

Goals of an IRB Review

The following points list the goals of IRB review and some comments on them:

1. *The detection of unauthorized research* is possible only if there is clear institutional policy demanding compliance with review board directives and severe sanctions for noncompliance. The full cooperation of departmental chairmen and research committee and the hospital or university administration is obviously essential and IRB's role in this endeavor is necessarily limited.

2. *Compliance with IRB directives* can be documented by properly executed consent forms, but the existence of such a form is superficial protection. Reviewers should be concerned with the personnel obtaining and witnessing consent. When a patient's physician is also the investigator, a third party obtaining consent may offer some protection against undue recruitment pressure, whether overt or subtle. A concern that the time of enrollment of a subject not precede the date of institutional committee approval is obvious and gives reviewers a sense of investigator dedication to compliance. The period between submission of a study to a review board and its eventual approval is a period of immense temptation for eager investigators who are anxious to initiate their projects.

3. *The evaluation of the subjects' exposure to harm* is required. All investigators have an obligation to promptly report unanticipated problems involving risks to subjects or others to their IRB. Their problems include adverse or unexpected reactions to biologicals, drugs, radioisotope-labeled drugs, or medical devices. To facilitate reporting, a research incident report (Appendix 13) should be filed by the investigator and the review subcommittee should evaluate the event to decide further actions. This includes reassessment of risk/benefit factors in the light of such occurrences.

4. *The assessment of subject awareness and involvement* in the research process is difficult if not impossible. Varying levels of education, understanding, interest, and anxiety make the evaluation of information given to subjects at the time of enrollment extremely difficult. The sensitive nature of a diagnosis or prognosis, the confidentiality of certain sexual, psychiatric, or drug-related studies, and the involvement of reviewers as third parties in a doctor–patient relationship are unavoidable complications hampering this evaluation. It is clear from the studies of Robinson and Meraz (1976) and Gray (1975) that patients recall minimal amounts from interviews conducted before open heart surgery. If such a result was found after giving consent to such an operation, it is unlikely that the description of an experiment protocol, its risks, benefits, and alternative therapies would be recalled more accurately.

The use of laymen in designing and assessing consent forms, the offer of a disinterested ear for problems, the clear statement of the subject's right to withdraw from the study at any time, and the guarantee that such withdrawal would not prejudice subsequent care of his condition are all important safeguards. Documentation, either through signed consent forms or follow-up interviews, of the subject's awareness of these rights is the ultimate goal of an audit, although it may be unobtainable in most instances.

Government and the Audit of Human Research

IRBs are required to review all approved clinical investigations yearly until such projects are concluded to assure compliance with directives of the IRB, and with federal law and its implementing regulations. Suspension or termination of approval of a clinical investigation may occur on grounds of noncompliance or where unexpected serious harm to subjects has occurred. Written reports to investigators, appropriate institutional officials, and the FDA stating reasons for suspension or termination are mandated.

An IRB has the right to observe the consent process or the clinical investigation with disinterested personnel if it deems such action to be appropriate. IRBs are required to report to appropriate institutional officials and the FDA any episodes of serious or continuing noncompliance. Research records shall be retained for at least five years after the completion of a clinical investigation.

Finally, at the first review following adoption of the new federal regulations, reviewers must determine whether the new regulations with their expanded elements of informed consent require changes in the consent forms of ongoing projects. If such changes are required, it is left to the IRB to determine whether revised informed consent should be obtained from subjects previously enrolled and continuing in the study. Grounds for requiring revised consents may include exculpatory language, failure to reveal risks, and failure to reveal the experimental nature of the investigation.

Practical Considerations

In these times of limited resources for colleges, universities, and hospitals, financial constraints force IRBs to produce maximal protection of human subjects frequently with limited fiscal resources. Since proposed federal regulations allow for the exercise of such discretion, varying levels of accountability and surveillance are possible. Minimally, investigator interviews and assurance of compliance can be obtained. Surveillance of research records and consent forms by the staff of the institutional review committee can be accomplished by sampling or complete review. A sample of subjects can also be interviewed to confirm research records and assess both their understanding of the research and their awareness of their rights as subjects.

The budgeting of limited institutional resources requires that the review procedure fit the magnitude of risk involved in the project. Intermediate situations, where risks, although present, are not severe, should require moderate expenditure of resources. In situations where potential risks of harm to patients are more than minimal, more attention is required to maximize patient protection.

The decisions as to the level and frequency of review that is required should be made at the time of approval by the review board. Investigator reports of unexplained side effects, alarming adverse reactions, morbidity, and mortality should always trigger an appraisal of the review mechanism selected for a study.

Review Mechanisms

With these objectives and considerations in mind, at the time of approval of a study, an IRB should make a judgment of the potential risks to subjects and the rate of accrual of subjects and select the optimal time for review. All active studies must be reviewed yearly, but those that involve more than minimal risk may merit more frequent attention. Investigators should receive written notification that the IRB has approved the study and should be informed of the review schedule. Usually investigators are instructed to notify the IRB chairman or staff when a specific number of subjects has been recruited. This notification initiates the review.

The IRB's staff contacts the investigator to conduct a preliminary assessment of activity of the study and adherence to institutional policy and directives by review of subjects' charts and consent forms. A meeting of the investigator and the reviewers is arranged. The review team usually consists of a physician, a lay member, and the staff member to the board, all of whom should be knowledgeable by having participated in the original review and approval process.

A review should attempt to document that:

1. Induction of subjects occurred only after all appropriate approvals were obtained.
2. Informed consent had been obtained in the manner prescribed by the IRB.
3. Only the subjects meeting with the approved entry criteria were enrolled in the study.
4. All side effects, deleterious results, or unusual incidents that accrued to the subject from his participation in the study were properly and promptly reported to the review board and dealt with in an effective and responsible manner by the investigator and his staff.
5. The course of the research had not changed without proper approval by the IRB.
6. Subjects were not exposed to risks other than those described in the original protocol. This is where an assessment of preliminary results will elucidate the continuing appropriateness and candor of the original consent form.
7. The rights and welfare of subjects have been adequately protected.

The reviewers probe as deeply as is deemed appropriate from evidence obtained, including interviews of subjects where they are felt to be of value. A report is drafted by the reviewers (Appendix 10), and the investigator is given the opportunity to comment on this report. All reviewers' findings of problems, noncompliance, etc., are presented to the full IRB for action. The full committee may recommend continued approval, changes in the consent form, changes in experimental design, or suspension or termination of the study based on its finding. No study should continue unless the board votes its approval.

Review Findings

The great majority of reviews reveal that investigators have followed review board directives. Some problems, however, ranging from minor irregularities to more serious violations have been uncovered through the audit procedure. Some of the most commonly encountered situations include the following:

1. The investigator has assigned one title to a grant proposal, another title to the document that received official IRB approval, and, sometimes, a combination of the two on the consent form signed by the subject. While all aspects of the research are otherwise identical to the materials originally submitted for IRB review, the discrepancies in titles are confusing and sometimes misleading. The researcher should be instructed to change all titles to conform to the IRB-approved document. Should he wish to change the wording, this request can be submitted to the board for approval.

2. The researcher has made some minor modifications in methodology, recruitment procedures, etc., or has altered some language in the consent form itself without review board approval. In hospitals, one might find dosage schedules altered. Investigators are instructed to submit all such changes to the review board; any revisions or modifications must be approved and added to the researcher's file maintained by the staff to the board. To encourage compliance with these regulations, an expeditious procedure for investigators wishing approval for such modifications is suggested. Upon receipt of the new materials, staff should submit them to the chairman of the subcommittee responsible for the original review and to the chairman of the IRB. When both have given written approval, the modifications should appear on the agenda for full-board review under a section entitled "Changes in protocols already approved." The approval should be recorded in the minutes.

3. The individual whose signature appears on the consent form as a witness is also shown as a co-investigator in the proposal. Since this might be construed as a conflict of interest, the investigator should be instructed to have a disinterested party act as witness to the consent process.

4. The consent form is not filled out in its entirety, for example, if either the date or the signature of a witness is missing. While these omissions may appear minor, potential legal repercussions are serious. The inclusion of a date signifies that the researcher did, in fact, recruit the subject after having received IRB approval. The signature of a witness should bear testimony to the fact that the subject read the consent form before signing it and had an opportunity to have questions answered.

More serious in scope and less frequently seen are situations where the investigator recruited subjects before receiving review board approval. The statements "I just wanted to test the waters" or "This is a preliminary pilot project" are totally unacceptable. Such practice is a violation of institutional policy, the policies of the IRB federal regulations, and research ethics.

Conclusions

The continuing review offers the IRB an opportunity to weigh its impact on its campus or medical center. The appearance of a number of ongoing research projects involving human subjects that have not been reviewed and approved by the IRB should spur the board to reassess its procedures as well as to develop education programs for their professional colleagues.

The most recent HHS and FDA regulations place greater emphasis on the board's responsibility to review approved projects. The schedule for review should be based on the degree of risk involved in the project rather than a rigid six or twelve month rereview rule.

In summary, the mandate for continuing review of approved research is an intrinsic part of the IRB's mandate to protect subjects. If the procedures for continuing review are constructed optimally, both subject and researcher will benefit substantially.

References

Beecher, H. K., 1966, Ethics and clinical research, *N. Engl. J. Med.* **274**:135–136.

Gray, B. H., 1975, Human subjects in medical experimentation, in: *A Sociological Study of the Conduct and Regulation of Clinical Research,* John Wiley; New York.

Gray, B. H., 1979, Institutional review boards as an instrument of assessment: Research involving human subjects in the U.S., a paper prepared for the Conference on the Assessment of Science, University of Bielefeld, West Germany, May, 1978. Reprinted in: *The Role and Function of Institutional Review Boards and the Protection of Human Subjects,* Public Responsibility in Medicine and Research, Inc., Boston.

Heath, E., 1979, The IRBs monitoring function: Four concepts of monitoring, *IRB: A Review of Human Subjects Research* **1**(5):1–3.

Levine, R. J., 1979, The evolution of regulations on research with human subjects, in: *The Role and Function of Institutional Review Boards and the Protection of Human Subjects,* Public Responsibility in Medicine and Research, Boston.

Robinson, G., and Meraz, A., 1976, Informed consent—recall by patients tested postoperatively, *Ann. Thorac.* **22**(3):209–212.

SPECIAL PROBLEM AREAS

Studies Involving Children

Gwen O'Sullivan

The two principal factors which set apart the review of research involving children from studies that use competent adults are the difficulties of obtaining a truly informed consent and the method for assessment of risk/benefit ratio. Since standard procedures for IRB review of research have already been covered in a prior chapter, this chapter will deal solely with those added dimensions regarding the use of minors. Children are defined as persons who have not yet attained the legal age of consent, which is likely to be 18 in most states but can vary widely.

In reviewing studies with children, it is first necessary for the IRB to fulfill all the requirements for review of any research project, such as ensuring that the research methods are appropriate to the aims, that the investigators are competent, that the criteria for subject selection are appropriate, that confidentiality of data is maintained, and that risks are minimized by using appropriate safeguards. Beyond these normal review requirements, certain special procedures are required specifically for research using minors.[1] The first of these is an assurance that studies have been conducted first on animals, adult humans, or older children where appropriate, before involving younger children. An IRB should always call on an investigator to provide data from previous studies except in those instances in which studies are not appropriate. For example, if a new drug's use is intended solely for a neonatal condition or one existing only in young children, it would be neither reasonable nor useful to require prior studies on adults. It might, however, be helpful to seek data from research done with immature animals or a like population to the one being studied.

Second, review of studies involving children requires provision for involvement of a parent, guardian, or advocate in the conduct of the research, partic-

Gwen O'Sullivan • The Children's Hospital Medical Center, Boston, Massachusetts 02115.

ularly when very young or handicapped children who are not able to consent or assent are involved, or when the research entails more than minimal risk. In addition to requiring adequate consent procedures, which include both parent and child, a review board may wish to recommend that the parent be present during the conduct of the research procedure. Clearly, this is not always feasible, as in the case of an experimental study added to a diagnostic cardiac catheterization or surgical procedure (owing to restrictions on access to operating rooms and special "laboratories") or for certain studies done during a long hospitalization where a parent is not constantly present. But in many instances, such as outpatient studies, psychological testing, or studies taking place in emergency rooms, the reassuring presence of a parent is logistically possible and can add to the child's tranquility and willingness to participate. The IRB at Boston's Children's Hospital has at times required the involvement of the parent in studies using normal controls or other research in which there is no direct benefit to the child. This requirement is felt to represent an additional safeguard to alleviate the child's anxiety and to offer assurance to the parent that the procedure is indeed being performed as described.

Third, a system must be developed by which the children themselves can agree to participate in a research procedure based on age, maturity, intellectual capability, or a combination of these, along with the consent of a parent or guardian. This issue will be dealt with later in this chapter.

Assessing the Risk/Benefit Ratio

Since children are considered a vulnerable population, the National Commission for the Protection of Human Subjects of Biomedical and Behavioral Research and, subsequently, the Department of Health and Human Services have proposed a specific system of risk assessment for these subjects. The FDA's proposed rules regarding children in research published in the *Federal Register* April 24, 1979 follow closely, if not exactly, the HHS guidelines in all their provisions.[2] There are two overriding concerns about research involving children. The first is research that is not clearly for the direct benefit of the child; the second is the concern about the child's and parent's consent to procedures involving risk and willingness to participate in nonbeneficial research. As presently defined by HHS and FDA, minimal risk means the "probability and magnitude of physical or psychological harm that is normally encountered in the daily lives or in the routine medical, dental or psychological examination of healthy children."[3] Risk/benefit assessment is central to both FDA- and HHS-proposed regulation. Each federal agency has defined four classes of research. The IRB has authority to review and approve the first three, but the fourth, which covers cases where the risk is serious and of indirect benefit to the child, requires the approval of both the IRB and HHS or FDA, whichever agency's rules would apply.

1. Research Involving only Minimal Risk.

If the IRB is of the opinion that the study under review does not place a child at greater risk than the type defined above, it may approve the research without detailed evaluation of whether the benefit is direct to the subject or of potential benefit to society at large. Most IRBs would tend to include here venipuncture, standard psychological and educational testing, behavioral observations, comparative studies of standard approved drugs, research on child motor and cognitive development, skin biopsies, echocardiograms, EEG, EKG, allergy scratch tests, urine collection, or any other procedures similar to those used by physicians as diagnostic measures.

2. Research Involving Greater than Minimal Risk but Presenting the Prospect of Direct Benefit to the Individual Subjects.

Certain studies belonging to this category are clear-cut and easy to define. An example might be the use of a new drug for which side effects are not expected to be severe, but whose investigational nature suggests that the occurrence of unanticipated or unknown adverse effects cannot be predicted. Interviews which raise sensitive issues or pose invasive or embarrassing questions but which could yield information of possible benefit to the child might also be placed in this category. Bone marrow, liver, kidney, or other organ biopsies would seem to belong here. When taking lung biopsy for research purposes during the course of cardiac surgery, the risk may be judged more than minimal, but potential benefit to the subject in assessing the child's condition and future prognosis might be considered of sufficient magnitude to justify the risk.

An additional point for an IRB to consider in judging research of more than minimal risk is the experience of its institution. If hundreds of such procedures have been performed with no or few complications, then an IRB has a more solid basis upon which to make judgements. One might also consider including such statistical information on the consent forms so that the subject and parents have a point of reference regarding the procedure.

The administration of combinations of investigational drugs for cancer is usually a high-risk procedure involving severe, possibly adverse effects but is one that offers the prospect of possible benefit to the patient. Such studies fall into this category. In fact, it would seem that all experimental measures, investigational drugs, or medical devices used in life-threatening situations would come under this heading. In these cases, with parental consent and agreement of the child when possible, one could view the risks of the experimental treatment as being of less significance that the risks to the child of the disease itself. However, members of the Boston Children's Hospital IRB frequently agonize (as I suspect do many others) over the review of such studies; concern is often raised about the ethics of approving last-ditch therapies which can cause more suffering than the disease, even though the potential benefit is to save a life. In practice, though, committees are reluctant to disapprove these protocols on the basis that the new drug should at least be made available to patients and the option given to them or the parents to accept or refuse further treatment.

In a middle group lie the types of research in which the risk/benefit assessment is less easily defined. There is an intermediate gray zone between benign procedures, such as blood withdrawal and dental exams, and the potentially life-saving research just mentioned. These are often the cases on which a review board must spend most of its time. Generally speaking, when the prospect of direct benefit to the child seems quite likely, a higher degree of potential risk, disturbance, or discomfort may be allowed.

It has been argued that the HHS definition of minimal risk is actually no risk at all since it is likened to the risks we all have to take in our normal daily lives, which presumably include crossing streets or driving an automobile. Representatives of the Office for the Protection from Research Risks (OPRR) at NIH have frequently stated that the wording of the regulations is open to individual interpretation and that institutions are meant to use the rules as guidelines in reaching decisions that best suit their individual needs and conditions.[4] Therefore, it appears both legal and logical for the IRBs to apply these provisions for risk/benefit assessment to the individual cases presented to them and to interpret them as the members of the board feel appropriate for their own research center. This would mean that procedures such as skin biopsies and spinal taps, use of such new devices as catheters and IV lines, use of investigational nuclear medicine isotopes whose radiation dose is the same or less than the standard isotope, or experimental measures added on to cardiac catheterizations could go into category 1 or 2 based on the board's experience with its own patient population and investigators.

The IRB is also called upon to assess potential benefit, which may be an easier task than evaluating degree of risk. Here, obviously, the key words are "potential" or "prospective" direct benefit to the child. Even risks inherent in clinical trials of new vaccines may be warranted for children with serious underlying illnesses whose treatment causes them to be immunosuppressed and, therefore, highly susceptible to infections. An example is the use of a new vaccine against herpes zoster or chicken pox in children who are immunosuppressed as a result of leukemia chemotherapy and are therefore at high risk to become infected. Clearly, the same trial for a vaccine against chicken pox in the general population of children would undergo a more rigorous review (see category 4).

3. Research Involving Greater than Minimal Risk with no Prospect of Direct Benefit to Individual Subjects, but Likely to Yield Generalizable Knowledge about the Subjects' Disorder or Condition.

In order for such research to be approvable, the board must find that "(a) the risk represents only a minor increase over minimal risk; (b) the intervention or procedure presents experiences to subjects that are reasonably commensurate with those inherent in their actual or expected medical, dental, psychological, social or educational situations; (c) the procedure is likely to yield generalizable knowledge about the subjects' disorder which is of vital importance for the understanding of the disorder." [5] Experience indicates that research in this category is sparse. Again, an individual review board's experience with its investigators and knowledge of its subject population must be taken into consideration. Given the option to interpret the regulations as best suits the institution's needs, the key point once more is for a board to decide how it will define "minimal risk" or "greater than minimal risk." If an IRB decides to use the standard that certain procedures, such as blood drawing, really pose no risk, then the next higher category, "a minor increase over minimal risk," could include a number of acceptable procedures. If, on the other hand, a board chooses to adopt a more stringent interpretation, a larger number of studies involving indirect benefit or benefit to future patients would not be approved.

Pediatric institutions appear to be more flexible in their use of these definitions. They have been dealing solely with research involving children for years, and their positive experience with investigators and subjects plus vast numbers of studies reviewed, have given them confidence to make decisions of a less rigid nature than institutions whose principal population is adult. It would be helpful for such centers to seek advice from IRB staff of pediatric institutions when dealing with problems arising from a protocol for children. A case in point involved the IRB of a general hospital in the Boston area which wrestled for weeks with review of a study involving children. Unable to resolve the issue of lack of direct benefit to the subjects, the chairman called on IRB members from pediatric hospitals for consultation; an ad hoc committee was formed which discussed the issue and advised the original IRB of its deliberations. Interestingly, the persons with more experience with pediatric research perceived the risk as negligible and would have approved the study even though benefit was indirect.

4. Research not Otherwise Approvable That Presents an Opportunity to Understand, Prevent, or Alleviate a Serious Problem Affecting the Health and Welfare of Children.

The proposed regulations in this area require that after fulfilling all other conditions for review of research involving children, the IRB must notify the Secretary of Health and Human Services who will consult with a panel of experts in pertinent disciplines and call for public comment on the issue, before approval can be given for research in this category.[6]To date, we know of no instances in which these measures have been used, but one can imagine that, had these regulations existed at the time of the Salk vaccine trials, such a panel of experts would have been assembled and review would have been a tedious and lengthy process.

Although no guidelines currently exist *per se* on the use of normal healthy children as research subjects, it would appear that this group would fall into either category 1 or 4. It is likely most IRBs would be opposed to approving studies which present more than minimal risk to healthy children except in special circumstances, such as investigations with siblings of affected children. These could range in degree of risk from experimental bone marrow transplants to use of psychological questionnaires. Bone marrow transplants from healthy siblings to children with leukemia, aplastic anemia, and other blood disorders are currently being carried out at several centers in the United States. Because of the still experimental nature of these transplants, a legal procedure is in place in Massachusetts to protect and assure consent for the minor donor. Here the physicians must appear in court each time a transplant is scheduled; a guardian *ad litem* is appointed to act as consentor for the minor donor, the rationale being that the parents may not be emotionally capable of consenting to such a procedure for the healthy child when the recipient of the marrow is also their own child. There may be cause to question this method of consent; though it is an enormously time-consuming process for all concerned, it is currently considered the best available objective route.

There are, of course, less dramatic examples of research using normal control children. These range from benign studies such as educational testing in schools to blood tests on siblings of children with genetic disorders to endocrine studies on siblings of affected children to clinical trials of new vaccines. In minimal risk studies, it would seem apparent that most controlled projects, particularly with educational and psychological testing and blood drawing, would comfortably rest in this "no" or "minimal" risk group. A board should have no difficulty approving such research with appropriate safeguards of confidentiality.

Should category 4 (general welfare) become a final rule and no new provision be made in the regulations for use of controls, then the procedure for approval would be highly complicated. In reality, it is unlikely that an IRB acting alone would approve the conduct of a high or medium risk procedure on a child who would not benefit at all. This kind of research might be embodied in the aforementioned clinical trial of a new vaccine for infectious childhood diseases, in which case the testing would doubtless be under government auspices.

Consent–Assent

By HHS definition, children are persons who have not attained the legal age of consent to general medical care as determined under the applicable law of the jurisdiction in which the research will be conducted. [7]Although all states have classified adulthood, there do not appear to be any federal or state laws pertaining to minors consenting to participation in research. Therefore, the standards for consent to medical care are commonly used. Exceptions are made in many states for "emancipated minors," individuals who are considered to have reached majority through marriage, parenthood, financial independence, or other means, depending on state law.

The fact that an individual may not be able to give consent adds an extra dimension to research involving children. The biological or adoptive parent or guardian authorized by state or local law must give permission for a child's participation in research. Except in unusual circumstances, the consent of one parent is sufficient. The concept of "assent" has been suggested in the proposed regulations as a way to involve the child as well, thereby creating a two-fold consent process. The proposed HHS regulations on the use of children as research subjects call for the establishment of "adequate provisions" on the part of the IRB for the solicitation of assent when "children are capable of doing so."[8] This latter phrase is where the difficulty lies. When is a child capable of understanding the procedure enough to be able to give informed assent? The original National Commission recommendations proposed the age of seven as the age for requiring assent. However, the subsequent HHS proposals suggested setting no specific age, but recommended instead use of the individual child's maturity as the criterion. The IRB is given the mandate of determining at what point age and maturity allow a child to assent in conjunction with parental permission. This judgment may be made for all children under a particular research protocol, or on an individual basis.

Since the experience of Boston Children's Hospital is entirely in the pediatric field, our IRB and administration have deliberated at length over this issue. It is our opinion that a set age is a useless criterion for assent even within a given research protocol because children differ so markedly in maturity and intellectual capability. Therefore, whenever appropriate or feasible, all children (except, of course, infants) involved in the consent process are informed to the maximum degree possible about the research project and are given the right to refuse participation. The concept of assent is not a legal one; it implies a lower level of knowledge than consent. It is important for the child to know exactly what will happen, whether and how much it will hurt, how long it will last, and that he or she can stop at any time. When there is a conflict between parent and child regarding participation, it is recommended that the child not be used as a subject unless there is some clear potential benefit to the child. For example, should a researcher desire an extra blood sample not for the direct benefit of the child, but to conduct further lab studies, the child's refusal should be honored even though the parent might agree to the blood withdrawal. When, however, there is potential for direct benefit that is important to the health or well being of the child and is available only in a research context, the assent of the child need not be obtained, so long as the parent consents. Examples of such research might be the administration of an investigational drug for the treatment of a chronic disease or chemotherapy involving experimental drugs for cancer patients.

Children's Consent

The role of older children and adolescents in the consent process must be separated from that of young children. Teenagers often have a strong voice in agreeing to or refusing experimental therapies for their illnesses, and physicians should be willing to honor their decision. It seems the main distinction to be made here is between therapeutic procedures of more than minimal risk with the potential for direct benefit, and those that do not hold out any prospect of direct benefit.

The next question is, who makes the decision as to whether or not a child is of sufficient maturity to understand a research procedure and give assent? Here again, one would have to divide patients/subjects and studies into categories. In a medical setting where chronic diseases are treated and children come in regularly for check-ups, the doctor–patient relationship is of utmost importance in establishing trust. In most of these situations, the child's personal physician would be the one to recommend enrolling the child into a study since the doctor is most knowledgeable about the patient's ability to comprehend the new procedure. Discussions among physician, parent, and child would ensue during which the decision would be made about handling the assent of the patient. An important consideration in deciding on a child's ability to assent is his familiarity with the proposed measures. For example, children with cancer are very knowledgeable about bone marrow aspirations and the degree to which they are painful; other children may have no acquaintance with the procedure.

Another category of subject would be those hospitalized on a one-time basis, such as surgical patients, who may be approached by a researcher (after gaining permission from the primary doctor) who wishes to conduct a randomized study on postoperative pain medication, administer a questionnaire, or conduct an interview regarding the psychological effects of a hospital stay on a young child.

Still another group would be those asked to participate in studies after being seen in the emergency room or in outpatient clinics. A board may want to impose more stringent measures regarding the need for an additional person to be present to judge competency for assent. On the other hand, they may feel no such provision is necessary for children recruited through schools for participation in educational or psychological testing of a benign nature. In these cases we require teachers and school officials to give prior permission to conduct such studies.

It is left to the IRB to determine if and when a third party should be called upon to assure that both patient and parent have understood the study and are willing to participate. At Boston Children's Hospital, nurses or social workers are often asked to act either as witnesses to signatures on consent forms or as advocates who audit the exchange of information between researcher/physician and subject. Although some hospitals use patient advocates to act in a general capacity, we are not aware of any institution which has provision for professional advocates whose charge is to monitor research consent procedures. In the absence of such an individual, then the assessment as to when and whether a child is capable of assent is left to the parent and physician on an individual basis for each child and each research project.

The HHS proposals cited earlier call for the consent of both parents for research under risk category 3 and also under category 4, where the study would go before the Secreaty of Health and Human Services. Exceptions would be cases of incompetence or unavailability of one parent, or in which a single parent has legal custody of the child.[9]

The proposals also make provision for waiver of written consent in cases where the requirement might fail to protect the children, e.g., abused or neglected children. IRBs frequently may be confronted with this dilemma in reviewing studies on child abuse. Here, legal and ethical issues intertwine as the difficulty of maintaining confidentiality arises. In some states, e.g., Massachusetts, a citizen has a legal obligation to report suspected cases of child abuse, so an assurance of confidentiality cannot be given to parents when asking their consent to participate in interviews regarding personal information about child rearing. After many debates on the issue, the IRB at Boston Children's Hospital decided that the fairest way to handle the problem was to state on the consent form that the law exists, that records are subject to subpoena, that the investigator will maintain confidentiality of the study results to the degree possible within this framework, and that the investigator will have to comply with the law should the necessity arise. A compromise has been developed whereby the consent form states that the researcher will not report directly to the state, but will pass on the information to the primary physician or social worker in charge of the case.

The problem of increasing anxiety levels in parents due to the presence of sensitive and embarrassing topics in questionnaires about child-rearing or home environment is present in much research. It is important that parents be warned ahead of time by means of the consent form of the nature of the questions to be asked so that they have a clear option to refuse participation or at the least to skip any questions they prefer not to answer.

Although it is commonly understood that the consent form is merely a documentation of a verbal exchange between doctor/researcher and patient/subject, and of less significance than the actual conversation, provision for the signature of the child on the form helps to draw the child into the consent process. Therefore, consent forms for pediatric research should contain two printed signatures lines, one for the child and the other for the parent/guardian. Whenever reasonable, according to the judgment of those involved in the explanation of the study, the signature of the child should be obtained. One problem occasionally encountered is that of the neonate whose parents are not available to consent to any research measures because the infant has been rushed from another location to a specialized tertiary care center. For these emergency situations, a paragraph can be added to the consent form providing for telephone consent. This method of consent should be used solely in cases where a therapeutic intervention of direct benefit to the infant is available only in a research context. The investigator reads the consent form to the father or mother (who is probably herself still in the hospital) over the telephone, while a third party monitors the conversation on another line. If the parent consents, the auditor signs the form as a witness, and at the earliest possible visit of the parent to the hospital, he or she is asked to sign the consent form.

Therapeutic vs. NonTherapeutic Procedures

Some IRBs may find it easier to make judgments regarding the participation of children in research based on whether the measures used are considered "therapeutic" or "nontherapeutic." The same risk/benefit principle is applied, but instead of creating categories based on increments of risks, the principle is applied within the two categories of therapeutic and nontherapeutic procedures. This chapter has concentrated on guidelines for review based on the classification of research into four categories used by HHS and FDA. Therapeutic measures are those considered to contain potential benefit directly to the subject. Nontherapeutic procedures are of no benefit to the subject but may benefit the health and welfare of other children or those who suffer from a similar disorder, or may further add to basic medical knowledge.

An article published in the *British Medical Journal* of January 26, 1980 by a working party on the ethics of research on children in England carefully explains these concepts.[10] Within the realm of therapeutic research, review procedures and ethical principles do not usually differ from those which apply to adults. The British authors cite only one kind of experiment in which an ethical dilemma is likely to arise: that of the comparison of two therapies in a controlled trial. The first question a review board should ask is, is this research necessary? If an investigator is requesting permission to assess a standard form of treatment by comparing it to another form because he questions that form of management, or because no clinical trial has ever proved that it was indeed the best method of management, then the IRB should assure itself that the answer to the question has not already been reported in the relevant literature. The second question the IRB should raise is whether the design of the trial is such that a statistically significant result will emerge with the use of a minimal number of subjects in a minimum period. Since one set of children will receive what *may* eventually turn out to be an inferior therapy, it is ethically imperative that this question be answered in the affirmative. If a clinical trial is to be conducted in which a placebo is used, the IRB should insist that the investigator demonstrate that this is the only method for assessing effectiveness of the treatment, and that reference to randomization and the placebo is made on the consent form.

Assessing the risk/benefit ratio involved in nontherapeutic measures is more complicated and challenging. The British group has suggested the three following categories:

1. Procedures which are part of ordinary care of an infant or child or those which involve noninvasive procedures, such as collecting urine, feces, saliva, hair, and cord blood at birth.
2. Procedures involving invasive collection of samples—for example, blood, cerebrospinal fluid, or biopsy tissue—taken from a child undergoing treatment. These samples used for research may be additional amounts to that required on clinical grounds, or not an ordinary part of the child's treatment—for example, collection of biopsy material during a surgical operation. If nontherapeutic procedures of no or negligible risk, such as venipuncture or collection of extra surgical tissue for research purposes during the course of an operation, could be of benefit to future patients suffering from the same disorder, there should be no question that the research should be permitted. A more difficult situation to assess would be performing a renal biopsy for research purposes during an abdominal operation. The risk would be judged more than minimal, so the benefit to other patients would have to be large to justify it.
3. Procedures which are quite apart from the necessary care and treatment of a child. Examples given are blood sampling, passage of an esophageal tube for pressure recording, application of a face mask for respiration studies, needle biopsy of fat or skin, and X-ray or isotope studies.[11]

Within each category, distinctions are made as to type of risk (none, negligible, or minimal), and whether the research benefits the health and welfare of other children or those with a similar disorder or simply adds to basic biological knowledge. According to this system, a higher degree of risk is allowable to a subject undergoing a nontherapeutic procedure if the benefit is accrued to other people (children or adults) rather than solely to the advancement of medical knowledge.

An undue emphasis on risk assessment may seem to emerge from these pages owing to the fact that the National Commission, HHS and the FDA decided to organize their proposed rules according to categories of risk. This could be unfortunate if it causes review boards to dwell too heavily on the risk side of the risk/benefit scale. Because of this emphasis in the federal rules, or their own fear of "harming" a vulnerable population or inexperience with the issues, some IRBs become overly stringent when judging research involving children. It is therefore of great importance that as much weight be given to the potential benefits, even if indirect, as to the possible risks or discomforts. In many instances, research of future benefit to children (e.g., the study of the prevention and treatment of childhood diseases, causes of genetic conditions and birth defects or problems of premature infants) can be performed only on a population of children.

The experience of the Boston Children's Hospital IRB has been that when the level of possible harm, pain, or discomfort is worrisome, measures to safeguard or alleviate these risks (such as guarantees of confidentiality, and available antidotes) have been required of the investigator. When instituted, these recommendations have considerably lessened the likelihood and occurrence of such problems.

IRBs should be aware of the danger of overprotecting subjects termed 'incompetent' insofar as consent is concerned at the risk of depriving them of their autonomy and ability to make decisions about their own health and welfare. In summary, equal consideration should be given to assessing both the risks and the benefits to be gained by the children themselves through the continuation of research that has been carefully reviewed.

Reference Notes

1. HHS Protection of human subjects, Proposed regulations on research involving children, *Federal Register,* Volume 43, No. 141, p. 31786–31794 July 21 1978.
2. FDA Protection of human subjects, Proposed establishment of regulations, *Federal Register,* Volume 44, No. 80, pp 24106–24111 April 24 1979.
3. HHS proposed regulations, 46.403 (j) and FDA proposed regulations, 50.3 (t).
4. Office for the Protection from Research Risks, verbal communication.
5. HHS proposed regulations 46.407.
6. HHS proposed regulations 46.408.
7. HHS proposed regulations 46.403.
8. HHS proposed regulations 46.409 (a).
9. HHS proposed regulations 46.409 (c).
10. *Br. Med. J.,* Guidelines to aid ethical committees considering research involving children, No. 6209, p. 229, 26 January 1980.
11. *Br. Med. J.,* Guidelines to aid ethical committees considering research involving children, No. 6209, p. 230, 26 January 1980.

Research on the Therapy of Cancer
With Comment on IRB Review of Multiinstitutional Trials

DALE H. COWAN

Introduction

Major advances in cancer therapy are attributable to clinical trials. There are two types of clinical trials: prospective and nonprospective trials. In prospective trials, subjects are allocated to two or more groups. One group serves as a "control" and receives whatever therapy is considered to be "standard" for the disease being studied. Standard therapy may be treatment with one or more drugs or, alternatively, no treatment. For example, the therapy given to patients in the control group of a clinical trial testing new treatments for metastatic carcinoma of the colon (colon cancer that has spread to distant organs) consists of the drug 5-fluorouracil. In contrast, standard therapy for the control group of patients with carcinoma of the colon that has not spread to distant organs is observation with no drug treatment. The other group(s) in the trial receive(s) the treatment(s) being tested. These may be a novel dose schedule of drugs known to have some effectiveness against the cancer being studied or a new drug not previously tested in the particular cancer.

When properly designed, prospective clinical trials establish conditions in which the outcome of the alternative treatment programs can be assessed and

DALE H. COWAN ● Division of Hematology/Oncology, Department of Medicine, St. Luke's Hospital, Cleveland, Ohio 44104, and Department of Medicine, Case Western Reserve University, Cleveland, Ohio 44104.

compared using statistical analyses. Prospective clinical trials commonly have the additional feature of randomization wherein the therapies being compared are allocated among the subjects by a chance mechanism. Subjects participating in prospective randomized clinical trials have, therefore, equal opportunity of receiving the standard treatment or the treatment(s) being tested.

The other type of clinical trial is not prospectively controlled. Rather, all subjects of the trial are treated with the therapy being tested. The results of the treatment are compared to those observed in persons previously treated for the same disease whether or not the latter results were obtained in the course of organized trials. The individuals previously treated are termed historical controls. The results reported with the historical controls serve as the basis for assessing the relative efficacy of the treatment being tested. For example, a clinical trial may be proposed to test the effectiveness of a drug in treating patients with metastatic carcinoma of the kidney. Since all known therapy for this disease yields a response rate of 10% or less, it may be decided to treat all participants in the trial with the drug in question. The results using the test drug would then be compared against the results of previous experiences in the treatment of this disease.

Regardless of the method chosen, clinical trials present a number of interesting and difficult issues for IRBs responsible for reviewing and approving them. These issues relate to (i) the manner in which the norms for determining ethical conduct in clinical trials can be applied to specific trials, (ii) the nature of the institution, that is, whether, for example, it is a university-affiliated or a community hospital, and (iii) the origin of the proposed trial, that is, whether it emanates from a group of investigators who collaborate to form a national cooperative group or from an individual investigator.

The ethical norms applicable to clinical trials, as identified by Levine and Lebacqz (1979), are that (i) there should be good research design, (ii) there should be a favorable balance of benefits and harms, (iii) the investigator(s) should be competent, (iv) there should be informed consent, (v) subjects should be selected equitably, and (vi) there should be compensation for research-related injury. This chapter will discuss the issues mentioned above by focusing on the norms and how IRBs in different institutional settings can apply them to particular clinical trials proposed by national cooperative groups and by individual investigators. Although the discussion centers on clinical trials of cancer therapy, the considerations are equally applicable to clinical trials for other types of research in treatment.

Research Design

The norm for sound research design derives from three underlying ethical principles (Beauchamp and Childress, 1979). These are (i) beneficence, the requirement to do good, (ii) nonmaleficence, the duty to avoid causing harm, and (iii) respect for person, the requirement that one subject not be used as a means to attain another's ends. The norm is expressed in the Nuremberg and Helsinki codes (WMA, 1964)[1] and is explicit in the current federal regulations.[2]

It is acknowledged that IRBs are not established to provide rigorous peer review of the scientific merits of a proposed study,[3] and cannot be expected to do so. Peer review of the scientific merits of a study is ordinarily conducted by the agencies which sponsor and fund the research. IRBs are established to safeguard the welfare of the subjects of research. Nonetheless, the norm of sound research design and the statement of policy that flows from it mandate that IRBs review the scientific basis for proposed clinical trials and assess the scientific and statistical design of the trial. Information necessary for this review and assessment include (i) the results of animal studies and of previous clinical studies or experiences with humans, (ii) whether there are similar studies currently underway elsewhere, (iii) the scientific rationale for the study being proposed, and (iv) the statistical basis for constructing the trial.

As an example, let us assume that it is proposed to evaluate a new combination of drugs for the treatment of undifferentiated small cell carcinoma of the lung. An IRB reviewing such a proposal should ask whether the proposed combination has been tried out previously in human subjects and, if so, under what circumstances, what the results of those preliminary trials had been, and whether the proposed treatment program is being tested elsewhere. In addition, it should ask whether the proposed treatment program includes a combination of drugs that have demonstrated efficacy against small cell carcinoma of the lung and whether plans for radiotherapy are included in the proposed treatment trial and comport with the known patterns of spread of the cancer. Also pertinent is whether the proposed trial is constructed in a manner allowing for the accrual of sufficient numbers of patients and allocates subjects between the proposed treatment and the standard form of therapy in a manner that allows the investigators to draw conclusions regarding the relative effectiveness of the proposed treatment as against the standard treatment.

It is clear that a thorough analysis of these matters requires a high degree of medical and statistical expertise. It is likely, however, that many IRBs do not have members who possess the necessary background and expertise to judge these issues. This is particularly so in community hospitals that may be cooperating in clinical trials through cancer control programs. IRBs in community hospitals are often composed primarily of lay persons or physicians whose practices are in the fields of general surgery, family medicine, or general internal medicine, and who are not involved on a regular basis with cancer chemotherapy or with research. Given these circumstances, how can IRBs exercise their responsibilities for review of research design?

It is suggested that one of two approaches may be adopted. One approach, which may be particularly applicable to IRBs in community hospitals, is merely to accept the information included in the project proposal as providing an adequate basis for justifying the study and a suitable design for achieving its purposes. This approach might be acceptable in the case of clinical trials designed and executed under the sponsorship of one of the national cooperative groups. National cooperative clinical trials have had the benefit of review by outside parties, such as the National Cancer Institute (NCI) and the Food and Drug Administration (FDA), during their development and prior to their submission to IRBs for approval. For example, the hypothetical clinical trial dealing with a new treatment of undifferentiated small cell carcinoma of the lung may have been proposed by a national cooperative group. In the development of the proposal the group would have addressed itself to the specific issues of scientific rationale and statistical design utilizing the talents of experts with extensive training and experience in the treatment of the disease and in the statistics in planning clinical trials. Their decision regarding how to resolve the various questions would have been reviewed and, if necessary, modified by trained personnel at the NCI and the FDA. Hence, the proposal submitted to the IRB would have had the benefit of an analysis with respect to research design substantially more extensive and more sophisticated than that within the capacity of any individual institutional IRB. Given this set of circumstances, it might be appropriate for an IRB to adopt the decisions of the cooperative group as affirmed by the NCI and the FDA. It could be argued that a further review by an IRB lacking the expertise available at the national level would not further contribute to safeguarding the welfare of the subjects.

The approach described, however, would be unacceptable in the case of clinical trials proposed by an individual investigator which had not been exposed to review by outside agencies. In the case of such investigator-initiated proposals, IRBs serve a quasi-peer-review role in assessing the design of the trial. By so doing, the IRBs have a relatively more significant function in safeguarding the welfare of potential subjects than where proposals have had prior outside peer review. To omit this assessment simply because of practical problems attendant to its performance would be unjustified and would result in IRBs being out of compliance with the spirit and the letter of the federal regulations.

The second approach IRBs can take to review the scientific design of clinical trials is to solicit the opinion of outside consultants. This approach can be used by IRBs in both community and university-affiliated hospital settings. It would be applicable to clinical trials proposed by national cooperative groups as well as those proposed by individual investigators. The consultants would be asked to determine whether suitable information exists to justify the proposed trial, utilizing the line of inquiry described above, and whether the trial as designed can be expected to achieve the intended purposes. Thus, in the example cited above regarding treatment of small cell carcinoma of the lung, the consulting oncologist would advise the IRB regarding the state of the art with respect to the treatment of this disease and whether the proposed treatment trial is a reasonable one medically. The consulting statistician would advise the IRB regarding the statistical design. While the outside reviews would not be intended necessarily to ascertain whether the proposed trial is the best one that can be done, they could well lead to improvements in the research design that would benefit both subjects and investigators. The use of the consultants would thereby enable the IRB to compensate for the limitations of expertise on the part of the members and thereby to carry out their responsibilities for assuring satisfactory research design.

Balance of Benefits and Harms

The norm of a favorable balance of benefits and harms, like that of good research design, also rests on the principles of beneficence, nonmaleficence, and respect for person. The norms of balancing risks and benefits is expressed in all codes of ethics and is stated in the federal regulations as a specific duty of IRBs. Federal regulations state that IRBs must determine that "risks to subjects are reasonable in relation to anticipated benefits to the subjects and the importance of the knowledge to be gained."[4]

To balance the risks and benefits of a clinical trial, IRBs must consider the disease being treated, the specific details of each treatment regimen, and the manner and setting in which treatment will be administered and responses of patients monitored. To illustrate, let us consider a proposed clinical trial intended to study the effect of adjuvant chemotherapy in women with Stage II carcinoma of the breast. This stage of breast carcinoma is that which exists in women who have undergone surgery for removal of the primary site of disease (that is, a mastectomy) and who have been found to have cancer present in the lymph nodes under the arm on the same side as the affected breast.

The benefit of pursuing this trial is judged by reference to the natural history of the disease and the likelihood of the cancer recurring at a later time. It is also related to the estimated likelihood that the anticipated effects will be realized. In the example at hand, knowledge of the natural history of breast cancer tells us that women who have 4 or more axillary lymph nodes containing cancer at the time of their mastectomy have a significantly higher incidence of recurrence of the disease within 5 years after their surgery than individuals with 1–3 involved nodes. Similarly, individuals with 1–3 nodes positive for cancer have significantly higher rates of recurrent disease within 5 years than individuals who have no positive nodes. For these groups of patients, therefore, the anticipated benefit from administering adjuvant chemotherapy is directly related to the number of affected nodes.

To estimate the harms that potentially might arise during the course of the clinical trial, consideration must be paid to the nature and potential severity of specific side effects from drugs given to women receiving therapy and the possibility of earlier or more frequent recurrence of disease in women in control groups receiving no therapy. In this trial, the side effects would include such symptoms as nausea, vomiting, hair loss, abdominal distress, and reduced blood counts. In addition there would be generalized mild weakness and lack of energy that commonly occurs with chemotherapy and the time spent in receiving treatment.

The calculus of risks must also include considerations as to whether side effects can be anticipated, detected, and treated, including whether facilities exist to treat side effects in the event they occur. For example, in the clinical trials of treatment for Stage II breast cancer, it should be confirmed that blood counts will be checked before each treatment and that appropriate treatment is available in the event that severe depression of blood counts occurs. It is recognized that detailed information regarding medical issues involved in clinical trials will likely be known only to individuals with expertise in the area. IRBs will therefore have to determine from the description of how the proposed trial will be performed or from advice of consultant experts that the necessary and appropriate preventive and antidotal measures are intended to be utilized.

It is evident that balancing the harms against the benefits is in a sense trying to compare apples and oranges. However, the issue comes down to the question of whether the anticipated side effects and potential harm associated with either receiving treatment or being a member of a control group receiving no treatment is justified on the basis of the expected benefits to be attained. In the case of the patients with breast cancer and 4 or more positive nodes with a high likelihood of recurrent disease, the answer could readily be in the affirmative. In the patients with 1–3 positive nodes the answer might also be in the affirmative albeit with less certainty. By contrast, in patients with zero positive nodes, the answer might well be in the negative. In this latter group of patients, therefore, it might not be appropriate to proceed with the proposed study.

Allowing for the essential uncertainty that surrounds the calculations of risks and benefits, the task can be substantially simplified if there are complete descriptions of the pertinent information in the proposals submitted for review. Thus, the members of the IRBs should be able to make informed judgments regarding the relative risks and benefits of clinical trials from the information that is provided by the investigators. Although a consultant can facilitate the interpretation of the information, the need for outside guidance is less in balancing risks and benefits than assessing research design.

Competence of the Investigator

The norm that the investigator(s) should be competent is related to that of good research design and upholds the same three principles that underlie

that norm. Both the Nuremberg and the Helsinki codes state that research
" . . . should be conducted only by scientifically qualified persons."
It is evident that the IRBs must ascertain and certify that the investigators
responsible for conducting clinical trials are qualified by background and
experience to manage the disease entities being treated and the treatment reg-
imens being tested. This can be accomplished by confirming that the investigators have met the standards for competence established by national groups
for certifying qualified experts. For example, the attainment of board certification in the subspecialty of medical oncology may serve as an indication that
an investigator has demonstrated competence in oncologic medicine. Further,
IRBs ought to assure themselves that the investigators do in fact practice in
conformity with the standards of the specialty in which the investigators are
members. This can be a very delicate matter for inquiry and, particularly in
the situation of community hospitals, may be a matter that is not so readily
ascertainable since a substantial amount of a particular investigator's practice
may occur within his private office setting. Adequately to fulfill their obligations to safeguard the welfare of prospective subjects of clinical trials, however,
IRBs may have to inquire from others within the professional community
whether a particular person is qualified to act as an investigator. It should be
noted that this inquiry can be facilitated, in the case of clinical trials sponsored
by national cooperative groups, by asking the principal investigator at the local
institution responsible for supervising the trial in a particular community to
certify that the investigators charged with the responsibility of conducting the
trials are in fact competent to do so.

In addition to certifying that investigators possess the necessary medical
qualifications for conducting clinical trials, IRBs should ascertain that the
investigators manifest " . . . a high degree of professionalism necessary to care
for the subject" (Levine and Lebacqz, 1979, p. 730). This determination
requires an inquiry into the relationship of the investigator to the prospective
subject. It is recognized that an investigator may relate to a subject in a dual
capacity, i.e., as physician and as investigator. In the traditional physician–
patient relationship, the physician's primary concern is the patient's welfare.
The physician seeks to do that which is in the patient's best interest. The physician acts as the patient's friend or "advocate." In the investigator–subject
relationship, the investigator has a major interest in the furtherance of the
research enterprise. A potential conflict exists in which the pursuit of the goals
of the research may not serve fully the goal of promoting the patient's welfare.
Although the conduct of the research is ideally a cooperative venture between
investigator and subject, the physician acting as investigator has a potential
conflict of interest between his or her allegiance to the patient and to the goals
of the research.

As an example, let us consider the situation in which a physician is treating a patient for newly diagnosed, nonresectable, non-small-cell carcinoma of
the lung confined to the thorax. Assume that a clinical trial is underway to test

a new chemotherapeutic regimen for this disease. The physician knows that radiation therapy is standard treatment but that it generally provides only partial, short-lasting control. He may agree that the proposed chemotherapeutic regimen is at least a rationally based treatment that merits testing but knows that hitherto non-small-cell carcinoma of the lung has been relatively unresponsive to chemotherapy. In determining whether to proceed with radiotherapy or enlist his patient into the clinical trial, the physician has to balance his primary duty to act on behalf of his patient's best interest against his desire to contribute to the generation of new information regarding the treatment of this disease. A potential conflict may arise that affects the recruitment by the physician/investigator of the patient/subject into the clinical trial and the ability of the latter to exercise free and informed choice with respect to his or her participation in it.

IRBs therefore should inquire as to the sensitivity of investigators to the existence of potential conflicts of interest and to the manner in which these conflicts can be minimized or avoided so that the interest in the research does not override the interest of the patient/subjects.

Informed Consent

The norm for informed consent is based on all of the ethical principles discussed previously. It is well established in codes and regulations for the conduct of research. For example, the Nuremberg Code states that the voluntary consent of the human subject is absolutely essential. The Helsinki Code states that each potential subject must be adequately informed of the aims, methods, anticipated benefits and potential hazards of the study and the discomfort it may entail.

The information that shall be provided for prospective participants in human subjects research shall include the following:

1. A statement that the activity involves research.
2. An explanation of the scope, aims, and purposes of the research.
3. A description of any reasonably foreseeable risks or discomforts to the subjects.
4. A description of any benefits to the subject or to others which may be reasonably expected.
5. A disclosure of appropriate alternative procedures or courses of treatment, if any, that might be advantageous to the subject.
6. A statement that the subject will be notified of new information developing during the course of the research that might affect his participation in it.
7. A statement describing how confidentiality of records identifying the subjects will be maintained.

8. An offer to answer questions the subjects may have about the research or the subjects' own rights.
9. An explanation as to whether compensation and medical treatment are available if injury occurs in research involving more than minimal risks, and who shall be contacted if harm occurs.
10. A statement that participation is voluntary and refusal to participate will involve no penalty and will not prejudice the subject's right to receive continuing medical care.[5]

In addition to these basic elements of informed consent IRBs shall also require that information shall be provided, where indicated, to the effect that (i) the particular treatment or procedure being tested may involve risks to the subject which are currently unforeseeable; (ii) foreseeable circumstances may exist under which continued participation by the subject may be terminated by the investigator without regard to the subject's consent; and (iii) additional costs to the subject may result from participation in the research, and the consequences of a decision to withdraw and that significant findings which may influence a subject's continued participation will be related to the subject.[6]

In addition to the elements enumerated in the federal regulations, IRBs must consider whether consent forms should include the fact of randomization in the case of prospective randomized clinical trials. Numerous arguments have been made for and against disclosing to prospective subjects the fact that their treatment will be selected by a randomization procedure (Levine and Lebacqz, 1979, p. 738).

Those who feel that the fact of randomization need not be disclosed to prospective subjects argue that since the alternative treatments to be tested are not known to produce significantly different results and since the physician would have to make an arbitrary selection of one treatment or the other for a particular patient, notification that selection of treatment is by computer rather than by the patient's own physician does not provide additional protection for the subjects and is unnecessary. The response to this contention is that a subject's ability to exercise full autonomy over what will be done with his or her own body is best served by notifying the subject as to how the treatment will be selected and by whom, even if the selection process is equally arbitrary whatever process is used.

The weight of the arguments favors the notion that for consent to be fully informed subjects must be notified that their treatments will be allocated in a random manner, i.e., selected by a process other than the judgment of their own physician. The meaning of the concept of randomization and the fact that it will be the manner by which treatment is selected is therefore considered to be an important and integral part of informed consent for participation in randomized clinical trials.

Implicit in the elements that comprise informed consent for subjects participating in clinical trials in cancer therapy is that subjects will be notified of the fact that they do have cancer and the extent of its spread. Current bioethical thinking views this to be an essential act in order for patient/subjects to give legally effective informed consent. Telling patients with cancer that they have the disease represents a substantial departure from past medical practices. A carryover of these past practices may exist in both university-affiliated and community hospitals. The current practice in the United States is that informed consent to participate in clinical trials requires that patients be notified of their diagnosis. Accordingly, a statement regarding the diagnosis is required in consent forms for participation in clinical trials that are sponsored by national cooperative groups. It is of interest that other Western countries do not feel that it is necessary or even appropriate to inform patients of the diagnosis of cancer as part of the consent process.

The elements listed above which need be provided for consent to be informed must be expressed in a written consent form. Let us consider what a consent form might look like for a study designed to assess adjuvant chemotherapy of breast cancer. The consent form would begin by stating that the patient/subject has been diagnosed as having carcinoma of the breast and that it has been found in one or more lymph nodes under the arm on the side of the affected breast. The form would then state that depending upon the number of lymph nodes involved and the age of the patient (that is, whether the patient is premenopausal or postmenopausal), there is a varying likelihood that the cancer will recur at a later date. The precise manner in which this information is imparted is critical since it is desired to provide the patient with sufficient information regarding potential future risks of recurrent disease without at the same time unduly frightening the patient as to her future outlook.

The consent form would then explain that experience with chemotherapy, with the manner in which drugs work, and our understanding of how cancer cells behave when they are present in relatively few numbers as against the larger numbers present in large masses, suggest that it may be possible to eliminate the cancer cells that may be present in the patient. It would be explained that there is no way of knowing whether a particular patient does in fact have any remaining cancer cells after the surgery, but that if some are present, the possibility exists that treatment given promptly rather than at a later date might be more successful in eradicating those cells. It is necessary to indicate that this is the theory underlying the proposed clinical trial and that there is no way of knowing whether this theory will prove to be true until certain numbers of patients are treated and the results of treatment are compared to the results from patients not treated. The patient would, therefore, have to be notified of the uncertainty regarding whether treatment will in fact be beneficial and at the same time told that there is a theoretical basis for thinking it will be.

The patient would next have to be informed of the potential risks and benefits of treatment and nontreatment. There would have to be an explanation of the side effects of the drugs to be tested, of measures that would be taken to monitor or anticipate the development of these side effects, and of steps that would be taken to ameliorate or treat them. There would then be a statement of benefits that may potentially accrue to participants in the trial to the extent that they could be reasonably foreseen.

The next portion of the consent form would describe alternative approaches to those proposed in the trial. In our example, the patient would be notified that one alternative is to administer no chemotherapy. Another alternative is radiation therapy to the site of the mastectomy. The anticipated benefits and risks of no treatment and of radiation therapy to the chest would have to be outlined. This would include a statement of both immediate side effects and complications that might arise from radiation therapy as well as information regarding the ability of radiation therapy to delay local recurrence and extend survival.

The consent form would then state that if, during the course of the trial, it should become evident that adjuvant chemotherapy imparted a significant benefit to subjects in terms of delaying recurrence of disease, or conversely, that it was detrimental to the patient in terms of the side effects, this information would be imparted to the subject so that she could decide whether to continue to participate. It would also be stated that the investigator might elect to discontinue the patient's participation in the research in the event that information was generated indicating that continued participation was no longer appropriate or permissible.

A statement would be included at this point defining randomization and what it means to have treatment selected by a process of random allocation rather than by the patient's own physician. It would therefore be clearly stated that rather than the physician deciding whether or not the patient would be receiving treatment or no treatment, this decision would be made by an allocation process outside the control of the physician.

The consent form would state whether the patient would have to pay additional costs as a result of participating in the research and whether or not compensation and medical treatment would be available in the event injury occurred as a result of the clinical trial. Finally, the consent form would (i) state who is responsible for the investigation and whom to call in the event questions or problems arise during the course of the treatment, (ii) offer to answer any questions the subject might have, (iii) explain that participation is voluntary and may be terminated at any time by the patient/subject, and (iv) indicate the manner in which the records of the patient would be kept confidential.

It is evident that a consent form with all these elements will be a lengthy one. In fact, consent forms for participation in clinical trials often run to three or four single-spaced typewritten pages. However, the anecdotal experience of those involved in clinical trials is that the majority of patients appreciate the full explanation provided in the consent forms and that these explanations do aid patients and their families significantly in determining whether or not to participate. Although there was widespread concern that these detailed, extensive explanations would frighten patients and reduce the incidence of participation in clinical trials by prospective subjects, there are no data indicating that this has occurred. Rather, the evidence appears to be that the more fully informed patient is able to participate in a clinical trial in a more meaningful way thereby making the trial a cooperative venture between the patient/subject and the investigator. IRBs should therefore not be deterred from requiring that consent forms are truly informative and should include all of the elements described above.

Two additional questions remain regarding informed consent in clinical trials. The first question is who should prepare the consent forms. Although it may be argued that a lay person, such as a lawyer, might be able to take the information provided in the clinical trial protocol and cast it into a form that would be most readily understood by prospective subjects, it is suggested here that the investigator is in the better position to perform this task. The investigator is the one who is fully informed as to the various issues that pertain to the clinical trial. Accordingly, the investigator is potentially in the best position to express the necessary information in a manner that is comprehensible to lay persons. This requires that the investigator must be capable of explaining the issues involved in terms that are understandable to the nonphysician. IRBs should insist that, if an investigator wishes to have patients participate in clinical trials under his or her authority, he/she ought to be able to explain to prospective subjects precisely what is involved in terms that subjects can understand.

The second question is whether or not a physician should act in the dual capacity as physician and investigator with respect to his own patients. Depending upon the particular circumstances, IRBs may be satisfied in allowing physicians to enlist their own patients in clinical trials in which they serve as investigators and to act as the caring physician during the trial. Alternatively, IRBs may wish to require that a knowledgeable third party, for example, another physician familiar with the disease and its treatment, or a party whose concern for the patient is undiluted, such as a close family member, participate in the enlistment process and in a monitoring capacity throughout the duration of the trial. In most university-affiliated and community hospitals, the matter should be decided with reference to the particular clinical setting for the trial, the type of patient/subject involved, and the nature of the disease entity under study. For example, in a community hospital having no medical oncologist other than the physician/investigator, and where the prospective subjects are all private patients of the investigator, it may be appropriate that a patient advocate in the person of a family member or a member of the nursing service be present at the time of the consent proceeding. In a municipal hospital which is university-affiliated and has other medical oncologists on staff, it may be appropriate for one of the other staff members to be present at the time the consent is enlisted so that the coercive elements that some deem to be inherent in such settings may be minimized. IRBs have to review each proposed clinical trial and specify conditions in which informed consent will be obtained on a case by case basis.

Equitable Selection of Subjects

More than the other norms, the norm that subjects shall be selected equitably is based on the principle of justice (Beauchamp and Childress, 1979). The principle of justice requires that the burdens and benefits of research should be distributed fairly. Unless a particular disease entity afflicts only a specific subpopulation of the community, it is inappropriate for subjects of clinical trials to be drawn primarily from members of that group. For example, it would be inappropriate for a clinical trial of a proposed treatment regimen for carcinoma of the breast to be tested solely in lower socioeconomic groups who receive their medical care in municipal hospital clinics. Rather, the trial should include women from all socioeconomic groups and racial backgrounds. Thus, no single socioeconomic or minority group in the community should be asked to bear the burdens of research.

The concept was incorporated in recommendations made by the National Commission for the Protection of Subjects of Research in its reports dealing with research groups deemed to be vulnerable by virtue of their limited capacities to consent, such as children.[7] The recommendations comment that, "the burdens of participation in research should be equitably distributed among the segments of our society, no matter how large or small those burdens may be." The regulations that were promulgated with respect to research with children state " . . . selection of subjects will be in an equitable manner, avoiding overuse of any one group of children based solely upon administrative convenience or availability of a population."[8]

The pertinent issues that IRBs should ask investigators to address are (i) the patient population from which subjects will be selected, (ii) the basis or rationale for selecting subjects, and (iii) the precise manner or setting in which subjects will be selected.

IRBs should ascertain that selection will not be made exclusively or even primarily from patients whose dependence upon the institution is such as to cause them to be reluctant to decline to participate for fear of loss of benefits. IRBs situated at institutions such as Veterans Administration hospitals where the majority of patients constitute potentially vulnerable populations may have little opportunity to insist that subjects belonging to these groups not constitute the major source of participants in clinical trials. Under these circumstances the IRBs, acting on behalf of the potential patient/subjects, must certify that the setting for selecting subjects and obtaining informed consent is such as to minimize to the extent possible the coercive atmosphere that potentially exists.

Compensation for Injury

The norm of compensation for research-related injury is a recently developed norm for research involving human subjects. It relates to the principle of justice. It is not expressed in any of the existing codes. It was first articulated in the HHS *Secretary's Task Force on Compensation of Injured Research Subjects*.[9] That recommendation states that "human subjects who suffer physical, psychological or social injury in the course of research conducted or supported by the PHS should be compensated if (1) the injury is proximately caused by such research, and (2) the injury on balance exceeds that reasonably associated with such illness from which the subject may be suffering, as well as with treatment usually associated with such illness at the time the subject began participation in the research." The norm was incorporated in the interim final regulation in which HHS specified that the availability of compensation must be included as an element in the informed consent form.[10] As noted above, an explanation regarding the availability of compensation and medical treatment in the event of research related injury is a designated basic element of informed consent in current HHS and FDA regulations[11]

Given this new requirement, how can IRBs apply it to clinical trials? Let us answer this by considering the clinical trial presently underway assessing intensive chemotherapy in patients with carcinoma of the ovary. In this trial, half of the patients are randomly allocated to receive the standard therapy for ovarian carcinoma, which consists of a single drug taken orally. The other half of the patients are allocated to a treatment regimen that involves intravenous administration of a combination of several potent anticancer drugs. The various agents in the combination regimen have a greater potential for inducing severe decreases in blood counts, hair loss, and gastrointestinal complications than does the single oral agent that constitutes standard therapy. In addition, the agents in the combination treatment can cause potentially severe impairment of cardiac and renal function. Thus, the treatment being tested may produce complications similar to those occurring with standard treatment or from the disease itself but may also introduce additional risk factors.

IRBs called upon to review this trial can satisfy the norm for compensation for research-related injury in one of two ways. First, they might interpret the regulations that embody the norm strictly. They might merely ascertain that the consent form includes a statement regarding the availability of compensation without inquiring further as to whether the investigators or the institution should provide compensation. Since schemes for providing compensation are not readily available, this approach would not impose upon the investigators or the institution requirements which they would have difficulty meeting. It would, however, serve to alert prospective subjects to the issue of research-related injury and allow them to consider the loss from such injuries in their decision to participate in the trial.

Alternatively, IRBs might determine which potential complications might arise from the disease or the standard therapy and hence would not be compensable simply because they occurred in an individual participating in a clinical trial. The IRBs could then determine which complications, if any, could be attributed to the participation in the clinical trial. They might then recommend that the investigators and/or the institution signify a willingness to compensate subjects injured as a result of participation in the trial or to provide medical treatment without additional costs. This approach might very well increase the cost for doing research and might make the performance of the clinical trial unacceptable to the investigators and the institution. IRBs do not have the power to require that compensation be made available. Consequently a recommendation by the IRBs to this effect would not be enforceable. A recommendation, however, would alert the investigator and the institution to the potential existence of a compensable injury. To the extent that this would promote additional safeguards for preventing such injury, the patient/subject would be benefited.

It is important that the matter of compensation be kept in perspective by IRBs. There is a substantial body of expert opinion in the oncologic community to the effect that the best treatment is done in the course of clinical trials. The legitimate concern regarding the cost of research-related injuries to particular subjects of clinical trials should be balanced against the benefit arising from trials to individual subjects and to the classes of subjects of which the individuals are members. It is suggested that this can be best accomplished by IRBs ascertaining that the potential subjects of clinical trials be notified as to the availability of compensation for research related injuries and that they be similarly notified as to the standards that apply in determining whether injuries are in fact research-related and therefore compensable.

General Considerations

Two general issues arise with respect to the role of IRBs in reviewing clinical trials of cancer therapy. One is the role of lay members of the IRB. It has been suggested in the discussion above that there are scientific matters that are unfamiliar to lay members of the IRBs. These matters can be judged by the professional scientific members of IRBs and/or by consultants to the boards. Other matters, however, can be readily addressed by lay members. Lay members serve an important role in ascertaining that the elements of informed consent have been adequately provided, that subjects are selected on an equitable basis, and that the matter of compensation for research-related injuries is adequately addressed. These matters do not require scientific expertise and are precisely the issues in which lay members can reflect the values of the community. The role of the lay members is therefore complementary to that of the physician/scientist on the IRBs.

The second issue relates to the role of IRBs in reviewing trials sponsored by national cooperative groups. The scientific background which underlies these trials and their statistical design are often complex and not readily understood by persons, including physicians, not intimately associated with the field of cancer medicine. The role of IRBs in reviewing protocols arising from the national cooperative groups can be determined by considering the norms which define the conditions that should exist for the ethical conduct of clinical trials. These norms involve considerations that are scientific as well as ethical. The scientific matters can be approached by relying upon the validity, accuracy, and completeness of the scientific information included in the protocol submitted for approval. Alternatively, they can be assessed by consultants available to the IRB from within or without the institution. The issues that are more ethical in nature, such as informed consent and equitable selection of subjects, are directly within the province of IRBs. There is no need to treat proposed clinical trials sponsored by national cooperative groups differently from any other research involving human subjects. Ultimately, the responsibility for safeguarding the rights and welfare of prospective subjects of research conducted within an institution lies with that institution's IRB. That responsibility cannot be deferred or abnegated simply because the source of a proposed clinical trial is a national group that exists outside the institution.

Reference Notes

1. *United States vs. Brandt, 2 Trials of War Criminals Before the Nuremberg Military Tribunals (The Medical Case)*, pp. 181–182, Military Tribunal I, 1947.
2. 45 CFR Part 46, Section 46.111(a)(1)(i), *Federal Register* January 26, 1981; and Part 56, Section 56.111(a)(1)(i), *Federal Register* January 27, 1981.
3. *Federal Register* preamble to HHS final regulations, January 26, 1981, p. 8377; and *Federal Register* preamble to FDA final regulations, January 27, 1981, p. 8967.
4. 45 CFR, Part 46, Section 46.111(a)(2), *Federal Register* January 26, 1981; and Part 56, Section 56.111(a)(2), *Federal Register* January 27, 1981.
5. 45 CFR, Part 46, Section 46.116, *Federal Register* January 26, 1981.
6. 45 CFR, Part 46, Section 46.116(b)(1)–(6), *Federal Register* January 26, 1981.
7. Research involving children, in: *The National Commission for the Protection of Human Subjects of Biomedical and Behavioral Research*, pp. 4–5, DHEW Publication No. (OS) 77-0004, Washington, D.C., 1977.
8. 43 CFR, 31793, Section 46.404(a)(6) *Federal Register* 1978.

9. *DHEW Secretary's Task Force on the Compensation of Injured Research Subjects*, 1977, DHEW Publication No. (OS) 77-003, Washington, D.C.
10. 43 CFR, 51.559, *Federal Register*, 1978.
11. Section 46.116(a)(6), *Federal Register* January 26, 1981, and Section 50.25(a)(6), *Federal Register* January 27, 1981.

References

Levine, R. J., and Lebacqz, K., 1979, Some ethical considerations in clinical trials, *Clin. Pharm. Ther.* **25**:728–741.

Beauchamp, T. L. and Childress, J. F., 1979, *Principles of Biomedical Ethics*, Oxford University Press, New York.

WMA, 1964, Human experimentation: Code of ethics of the World Medical Association, Declaration of Helsinki, *Brit. Med. J.* **2**:177.

13

Surgical Research

Myron E. Freund

Surgery today is safer, and the anesthetic risk is less, than at any previous time in history. Continued progress in improving the surgical treatment of disease and in reducing surgical and anesthetic morbidity and mortality can take place only in an atmosphere where responsible, carefully performed surgical research is encouraged. To the nonsurgeon (especially the layman) serving on an IRB, it may seem that surgical research *per se* presents a great magnitude of risk, since surgery itself, even without a research component, enjoins substantial intrinsic risk. The IRB must therefore commit itself to understanding the needs and problems of surgeons and judge surgical research projects in an appropriate context.

Surgical research involving human subjects is not substantially different from human research in other medical disciplines. The IRB and the researcher must be concerned with proper design of the protocol, assure adequate protection of the study group, and justify an acceptable risk/benefit ratio before a research proposal can be approved and undertaken. The basic principles enumerated elsewhere in this volume are equally applicable to the surgeon and his patient.

This chapter discusses the various areas that are particularly pertinent to the surgical community in carrying out research. It is hoped that this will provide a surgical perspective for IRB members and help them to be responsive to the needs and problems of surgeons undertaking research commitments. It is also intended for the research oriented surgeon, to help him understand the principles IRBs must apply in reviewing research proposals.

New operative procedures generally evolve gradually. Those involved in

Myron E. Freund ● Department of Surgery, North Shore Univeristy Hospital, Manhasset, New York 11030, and Department of Surgery, Cornell University Medical College, New York, New York 10021.

their development may be convinced that clinical trials would prove their advantages, if performed. This conviction is felt to justify proceeding directly from pilot studies to unrestricted clinical usage, on the untested premise that to deny broad usage would be an unethical denial of the best treatment. The utilization of randomized and/or cooperative controlled clinical trials in surgery must, therefore, be discussed as well as the need for surgeons to participate in research. The Food and Drug Administration's (FDA) monitoring of new devices, National Institutes of Health (NIH) guidelines of particular relevance to surgical research, the monitoring function and the use of surgical specimens for research purposes are also discussed.

The Need for Surgical Research

A report of the National Research Council Committee on a study of national needs for biomedical and behavioral research personnel recommended an increase in the number of physicians involved in clinical investigation as being in the national interest. David B. Skinner, in his 1979 presidential address to the Society of University Surgeons, delineated these needs as they apply to academic surgery. Between 1965 and 1974 the number of medical students increased by 55%, the number of interns and residents in academic medical centers increased by 87%, and the number of medical schools increased by 30%. This has created a substantial need for more academic surgeons to teach undergraduate- and graduate-level medical education. Unfilled but funded teaching positions in academic surgery in medical schools and teaching hospitals numbered 494 in 1972. In the past five years, recruitment of academic surgeons has substantially declined. There is serious concern over the development, recruitment, and retention of academic surgeons today.

The proportions of physicians funded by NIH research grants, acting as principal investigators for the first time, has declined from 44% in 1966 to 22% in 1975. A questionnaire answered by chief residents graduated from three leading teaching hospital surgical programs since 1965 showed that a full-time research experience in the military or as a research fellow in a teaching hospital resulted in a high proportion of surgeons choosing a career in surgery. Those who chose academic surgery published an average of 5.5 papers a year since entering academic practice. The higher the academic rank achieved, the greater the number of papers presented per year. Dr. Skinner concluded that a full-time research experience was the most important determinant in the selection of an academic career for many of the surgical residents polled.

Termination of financial support of many academic surgical training grants has substantially curtailed postgraduate surgical endeavors in research. If we are to recruit and prepare the next generation of academic surgeons, the importance of providing research opportunities is obvious. Sources of funds are continually being sought by leaders in academic surgery in an atmosphere of dwindling support.

Academic surgeons currently support themselves largely by their clinical practices and this takes time from research and teaching activities. In addition, full-time staff are expected to carry out administrative assignments and responsibilities which further reduce time for teaching and research. Another factor which mitigates against a choice of a career in research medicine is financial remuneration. The full-time academician in an academic setting are doing very similar tasks but the full-time salaried physician is generally less well recompensed. While the NIH does support surgical investigators to a limited extent, there has been a decrease in applications from surgeons for NIH support in recent years, accompanied by a generalized decrease in NIH support of new investigators, including surgeons.

Surgeons at all levels, academic or clinical, must be encouraged to develop special interests and expertise that would be appropriate for clinical research pursuits. Such expertise among clinicians will be of eventual benefit to their patients and improve their collective value as a national resource.

The Clinical Trial

Properly designed clinical trials of new surgical techniques would seem to be an obvious component of the contemporary approach to scientific medicine. Advances in biostatistics and computer science have put us in a position to substantially refine the data base from surgical research by utilizing well-conceived, randomized clinical trials. The data base developed from surgical research becomes the basis for clinical surgical decisions. Yet in spite of the need for expanded information, the randomized clinical trial is a divisive force in contemporary surgical spheres, primarily for ethical reasons. Extensive clinical trials that randomize procedures with solid statistical grounding free of bias are, nevertheless, a required method for refining that data base.

Advocates of a particular procedure are generally reluctant to participate in such randomized trials for a variety of reasons. Very few surgical residents, during their formal training experience, are exposed to randomized clinical studies. Randomized clinical trials often represent an experiment to the surgeon, as well as to the subjects, and there is a natural reluctance to become involved with the unknown. In today's medicolegal climate, where surgeons are prominent targets of litigation, there is reluctance to enrollment of patients in any form of clinical trial. It is axiomatic that a surgeon must always do what he knows to be best. Where there are comparable alternative operations available to remedy a particular problem, a surgeon's knowledge and experience with a procedure make it best "in his hands," because he may be less familiar with alternative operations. Those who teach surgery are not always experienced with alternatives to the surgical treatment of a particular disease because they have not personally used other techniques and may feel other approaches are suboptimal. For example, cancer of the breast can and is being treated with radical mastectomy, modified radical mastectomy, simple mastectomy, and lumpectomy; all of these are performed with or without radiation and/or chemotherapy. None of these procedures is experimental, and all have their strong advocates and supporters. Cooperative clinical trials are underway for the first time in an attempt to increase the data base for the management of breast cancer. Recent advances in plastic and reconstructive surgery have provided substantial improvement in the cosmetic result after radical mastectomy, which has made the procedure more palatable to some patients and their surgeons. This ancillary consideration has to influence the decision, since morbidity and quality of survival are affected. On the other hand, reconstructive surgery is expensive and time-consuming, and carries its own intrinsic risks. The surgeon who feels that one of the choices of surgical management for breast cancer is best will feel it his moral and his ethical obligation to tell his patient that his particular choice of therapy is the best for her situation, and to offer any other would not, in his opinion, give his patient the best opportunity of a good result. How can he then submit his patient to a randomized clinical trial, denying his patient what he feels to be the best form of treatment?

Dr. Robert E. Condon, president of the Central Surgical Association, in his presidential address to that society in March 1979, presented the opinion that the clinical trial, by adding to the body of scientific knowledge, helps the surgeon to make better therapeutic choices for future patients and, therefore, represents a better ethical decision. In this regard, it should be noted that much of what the surgeon assumes he knows is not based on solid scientific data, but rather on training, experience, and reinforcement. The choice of treatment is neither more or less likely to be correct if made arbitrarily than if assigned randomly in the clinical trial. The two courses of action can thus be considered ethically equivalent in terms of patient risk, since the alternatives within a properly designed clinical trial are all established and acceptable alternatives.

Dr. Condon also asks the question: with how much certainty need a surgeon hold his opinion? If an opinion, even a tentative one, about the relative superiority of a treatment approach compels the surgeon to act on that opinion, participation in a randomized evaluation of the treatment is not possible. Most surgeons have developed opinions about the controversies with which they must deal. It should be remembered, however, that even an opinion held with strong conviction is not a sufficient basis for ethical action; passionate conviction does not make an incorrect opinion into a correct one. The proposal of a clinical trial is based on the assumption that there is a lack of scientific knowledge about the worth of a particular technique. The answers to the questions posed by clinical trials are not available by any other means. That is the reason for clinical trials and for the surgeon's obligation to participate in them when possible.

Each surgeon modifies the technique involved in performing any surgical procedure with each patient, in order to accommodate the unique problem that confronts him. This modification is a necessary and appropriate action for a surgeon to take. These minor modifications may progress, and in a given setting they may ultimately lead to something that may be rather different from their "classical" point of origin.

While the resultant new technique may seem quite different or novel to the general medical and surgical community, in the eyes of those involved in its evolution there is no such feeling. A pilot study of several patients will have been performed with success, perhaps equaling or surpassing that of the conventional approach to therapy. The surgeons primarily involved and those who cooperate in testing the new procedure in a pilot study may be enthusiastic in their advocacy of the new procedure. To submit their patients to a large clinical randomized trial is not ethical in their opinion, since they have already seen that their new approach is better. Their assumption is that they would not be doing the best for their patient, and yet, in terms of statistically significant, unbiased, truly tested knowledge, a sound scientific data base does not exist. They are not fearful that their new procedure will fail the scrutiny of open and widespread trial. Rather, their concern is ethically for their patient being denied what they already believe to be superior. In discussing this dilemma, Dr. Condon has pointed out that such an ethical quandary is avoidable if the pilot study concept is modified and all new procedures are evaluated by being randomized from the beginning. Dr. Condon points out, however, that this is unlikely to develop because of the evolutionary nature of much surgical development.

Historically, surgery has adopted many techniques and procedures without the benefit of scientific evaluation, some of which have later been found to be hazardous or ineffective, e.g., splenectomy in childhood after relatively minor splenic injury, mastoidectomy for infected mastoids, and tonsillectomy for all children—all of which are done less frequently today. Any surgeon utilizing a new procedure is in essence conducting an uncontrolled experiment, without adequate records, controls, or explicit consent from the patient. Often such surgery is performed under conditions in which little or nothing can be learned. The technical capabilities of accomplished surgeons are not sufficient reason for uncontrolled experimentation to be condoned by the surgical community or the medical community in general. Clinical trials can produce reliable data. They negate the influence of bias. They are ethical because they expose fewer patients to the risks of ineffective treatment and provide a more solid foundation for surgical judgments. Surgeons think in terms of probability every day. The process of surgical judgment involves intuitive weighing of various probabilities in any given clinical situation. Apparently authoritative but frequently unsupported opinions of colleagues enter into this decision-making process. Since so much of surgical judgment and prognostication involves probabilities, the precise basis for controlled clinical trials, it would seem reasonable that surgeons should be enthusiastically in favor of clinical trials. But surgeons find themselves in an uncomfortable and unfamiliar role when suspending judgment and independent initiative while conforming to the restrictions of the trial protocol. The mechanics of randomization result in an uneasy feeling about the outcome, and many surgeons prefer to avoid participation for this reason. It is necessary to accept the limitation of individual experience as a guide to action, in order to accept the concept of the clinical trial. It is easier to do this, however, by recalling that reduction of ignorance is professional responsibility and should be accorded a high priority in surgical practice and surgical education.

The development of a new operative procedure generally begins with animal experimentation with an animal model that is as similar to human physiology as possible for the purposes of the study. This is followed by an appropriate pilot study in human subjects. A clinical trial that is properly designed and randomized and free of bias should ideally precede the introduction of a new procedure to general clinical advocacy. Ludbrook (1977) equated clinical trials of operative surgical procedures with the very essence of surgery. Based on a worldwide search of surgical literature from 1970 through 1973, he reported only 8 prospective clinical trials of surgical procedures that were free of bias. The clinical trial is used more broadly today, but its potential remains largely unexploited.

An example will perhaps illustrate some of the problems concerning the randomized surgical trial. The surgical treatment of massive obesity is an area of substantial controversy. Although there have been trials comparing operative procedures for obesity, there are scant data on surgical treatment compared to medical treatment and/or compared to no treatment at all. Such a study was undertaken in Denmark, and the results were published in *Lancet* (Andersen, 1979). The subjects in this study were massively obese individuals who had heard about the operation and presented themselves for evaluation. After medical information had been collected on each person, they were randomly allocated to a surgical or medical group by the investigators. *Consent for randomization was NOT obtained.* Patients allocated for medical treatment were *deceptively* told that their surgery could not be performed for one reason or another. After observing the patients in both groups for many months, the investigators concluded that the surgical treatment, although not without risks, provided greater weight loss and improvement in quality of life for such persons.

The ethical quandary produced by this study was commented upon in an accompanying editorial (1979), as well as elsewhere (Stollerman, 1980). The *Lancet* editorial board felt that the study was ethically unsound, but, curiously, they chose to print it anyway. It probably would not have been accepted in most American medical journals. The subjects comprising the medical control group were not only disappointed; they were deceived. A scientific bias in the study now becomes evident; the patients in the medical control group could not have been expected to follow a medical regimen with any degree of success, both because of their past record of failure, and because of their current disappointment at being rejected for surgery. (After all, the patients for the study were all self-referred and wanted the operation.) A favorable outcome for the surgical group was, therefore, assured from the outset. An American IRB faced with this type of study would be forced to reject it, not only because of the lack of informed consent concerning randomization, but also on the grounds of lack of scientific validity.

Cooperative Studies: Multicenter Trials

A number of interhospital cooperative studies involving surgical diseases have been undertaken during the past decade. For example, an interdisciplinary group was recently formed by the National Institute of Arthritis and Metabolic Diseases to study the risk/benefit ratios for new operative procedures involving digestive diseases. The committee included gastroenterologists, pediatricians, radiologists, and surgeons. The work of this group has stimulated continued interest in the collection of risk/benefit ratio data for new and improved procedures, and also has led to the development of improved methods of collecting data and of educating the medical community. Such pooling of data and the cooperation of multiple medical disciplines has ample precedent. National and international registries and procedures in certain diseases such as kidney transplantation have been functioning for approximately 15 years. Cooperative studies, such as this, increase the data base and enable a more rapid accumulation of statistically significant numbers of patients within a given protocol.

Each cooperating institution in a multicenter trial must accept full responsibility for the way the protocol is utilized and monitored. The IRB must satisfy itself that the principal investigator at the institution is satisfied with all aspects of the validity and design of the protocol with adequate recourse to the overall principal investigator so that if modification is desirable, input for discussion of revision is possible. There is a tendency when cooperating in a large study to capitulate responsibilities to the primary or originating institution's review board, since that board will have reviewed the protocol before it is submitted to cooperating institutions. Since cooperative studies tend to be prestigious, and the duplication of effort in the presence of a sometimes overburdened IRB schedule creates a temptation to avoid seemingly needless repetition, it is worth emphasizing the responsibility of each IRB and local principal investigator to the patients who participate in such cooperative studies. To safeguard the safety and rights of patients at a cooperating institution, all normal monitoring and consent mechanisms must be maintained, and full disclosure to the IRB of complications developing from the research must be accomplished. It may be a reasonable policy for the IRB to include the right to review all collective data for approval before publication on the part of each cooperating institution to assess the validity of the conclusions and to protect itself from erroneous, misleading, or controversial conclusions.

Risk/Benefit Ratio and Surgical Risk

Contemporary techniques in anesthesia and modern antibiotic therapy have substantially reduced the risk of surgery. Outstanding advances in vascular and cardiac surgery, joint replacement, and transplantation have gone hand in hand with breakthroughs in immunology, microsurgical techniques, and hyperalimentation. The benefits derived at this point, observed retrospectively, far outweigh the risks of the pioneering efforts in these various areas of outstanding achievement. In many instances the risk to the individual was extremely high in the evolutionary stages of these advances, but not when compared to the prognosis with known modalities of therapy that existed prior to these advances.

Although it is a difficult question to weigh, it is the responsibility of the IRB to consider objectively the cost to the patient in terms of quality and longevity of survival. Where current modalities do not offer a good prognosis, it is reasonable to accept substantial risk when the validity and design of a proposed new procedure seem to warrant it. Justifiable risks must be taken. Such risks are accepted in research involving new drugs used in cancer chemotherapy with the high toxicity and extensive morbidity commonly associated with the use of such drugs. The same liberal attitude is warranted in considering certain surgical protocols. Acceptable risk can be quite high when weighed against the current cost in survival quality and longevity. For example, colonic bypass of esophageal carcinoma cannot be compared to enteric bypass for obesity. The risk must be weighed against the threat of the disease and available alternative modalities of treatment.

Devices Used in Surgery

The FDA now regulates the use of new devices just as it does the use of new drugs. In this context the IRB is responsible for approval and monitoring of the use of new devices in surgery. Protocols involving their use must be approved by the IRB before they may be utilized.

For example, intraocular lens implantation is an accepted surgical procedure, not an experiment. This is a widely utilized and accepted modality of treatment for cataract. The lens itself, however, is under FDA trial status at the present time. The technique has evolved to the point where it is now being extended to use in children with congenital or traumatic cataracts. The IRBs in some institutions are being asked by ophthalmologists to determine whether this procedure should be performed in children. Since the use of this modality in children is new, it represents a new operative approach and is in that sense investigational. The risk to the institution is substantial in any given instance, if the operation fails, resulting in blindness to the child. While the likelihood of this catastrophe is small, the surgeon is asking the institution to share the risk by obtaining IRB approval. The FDA has chosen to make this an IRB problem by categorizing *the lens* as an experimental device, although the surgical technique is not. How does one weigh the risk and the potential benefit in such a circumstance? Where other nonsurgical modalities are perhaps safer, is this an instance where a controlled clinical trial is warranted? The decision is probably best made in consultation with the hospital's medical board, research committee, and legal counsel.

IRB Monitoring Function in Relation to Surgical Research

Appropriate periodic reporting of progress and all reporting of untoward events involved in a surgical protocol must be spelled out in advance in consultation with the investigator. In surgical research, as elsewhere, it is difficult to assure reasonable compliance using only written reports. Periodic attendance at IRB meetings by an investigator may allow an opportunity for open discussion of problems and progress. This will give the IRB a better appreciation of the investigation and of the investigator. The IRB should expect a full discussion of all untoward events, and the researcher should recognize this as part of his responsibility. It is beyond reasonable expectation for the IRB to provide a police function. The chairman of the department of surgery or his designee should be responsible for monitoring surgical research in the clinical setting, in the operating room, and on the surgical wards. In some institutions, high-risk research is monitored to see that research protocols are on the charts, and that those involved in implementing the protocols are adhering to them. This can be done to a great extent by trained nonmedical personnel. To carry this form of monitoring into the operating room, however, is difficult and inappropriate. The IRB must accept this limitation and the good faith of those in the surgical department delegated to carry this responsibility. Where the risk is great, the IRB may recommend that an ombudsman or third party be present at the time that the protocol is discussed and consent requested by the researcher to assure that the procedure and its risks are fully understood by the patient. There is a danger in overutilizing such a technique where the risk does not definitely justify it.

Surgical Specimens Used for Research

The need for appropriate consent has been well established in the use of cadaver organs for human transplantation (e.g., cornea and kidney). Where biopsy material is requested as part of an investigational project, it is imperative that the IRB satisfy itself that the scientific validity of obtaining such biopsy material justifies the risk and that the material is being obtained in a surgically acceptable manner. The question to be asked is whether such biopsy material truly enhances, in a significant way, the validity of the answers sought by the protocol. Is the patient fully informed of his rights and is the knowledge obtained from such biopsy material available to the patient and his physician in situations where it may be of potential benefit? This may be considered to be the subject's right and is not an unreasonable expectation.

In situations where biopsy material is obtained coincidental to an open operative procedure, but is not part of the operative procedure, it introduces an additional though generally minimal technique to the operation. For example, if material aspirated from a cyst or a segment of subcutaneous fat is being used for research purposes that do not add substantial risk to the operative procedure, there may be a temptation to bypass the mechanisms of protection that the IRB provides to the researcher, in terms of consent and full disclosure of the purposes and potential benefits and risks of such a biopsy. The emphasis here should be on the fact that the risk in negligible. The IRB must keep this degree of risk in mind when deciding how much to ask of the surgeon or how much the patient need be informed. There is a tendency to err on the side of requiring consent in all instances and to request more than is necessary in terms of compliance of the investigating surgeon. It is probably not necessary to obtain a consent in all such instances. The mechanism of expedited review for studies involving surgical specimens should be applied to these cases.

Conclusion

The IRB must be satisfied that the technical design of a research protocol in a surgical setting is satisfactory and that the question asked is reasonable, clear, and potentially answerable by the protocol. In addition, the validity of the concept must be satisfactory, i.e., that the question should be asked at this time in this way by this researcher. The cost in terms of longevity and quality of survival must be weighed in the risk/benefit equation, bearing in mind that there is an intrinsic risk involved in all surgical procedures, and that alternative modalities are not without substantial risk in some instances.

The particular project should be timely. If it is a pilot study, have adequate animal studies been performed where appropriate? If adequate animal studies and a pilot study have been performed, is the clinical randomized trial of sufficient scope to provide the data base necessary to determine the efficacy of the new operation? Is the institution a suitable place for the particular research project to be undertaken and by the particular investigator? Are federal and local governmental regulations and guidelines being followed?

If surgical knowledge is to be advanced, then the IRB can participate in a constructive way in encouraging and supporting appropriate and even high risk research, but must do so with appropriate safeguards. There is a great reluctance on the part of many surgeons to become involved in research because of the formidable bureaucracy that must be encountered. The IRB should be supportive and helpful to the surgical investigator, while fulfilling its responsibilities to the patient and the institution.

References

Andersen, B., 1979, Randomized trial of jejunoileal bypass versus medical treatment in morbid obesity, *Lancet* **2**(8155):1255–1258.
Editorial, 1979, Disinterested intellectual curiosity among surgeons, *Lancet* **2**(8155):253.
Ludbrook, J., 1977, Is surgery a scientific discipline? *Aust. N. Z. J. Surg.* **47**(6):732–773.
Stollerman, G., 1980, Randomizing ileo bypass for morbid obesity—Willy-nilly, *Hospital Practice,* **2**(8):35.

Bibliography

Adelman, S., 1978, Research hinders clinical training in surgery (letter), *N. Engl. J. Med.* **299**(2):102.
Condon, R. E., 1979, Improving data base—Whose obligation? *Surgery* **86**(3):363–367.
Eisman, B., 1978, Epidemiology of surgical research leadership, *Ann. Chir. Gynaecol.* **67**(4):129–133.
Katz, A. E., 1978, Research does not hinder clinical training in surgery (letter), *N. Engl. J. Med.* **299**(17):961.
Skinner, D. B., 1977, Interactions of federal programs with academic surgery, *J. Surg. Res.* **22**(4):429–434.
Skinner, D. B., 1979, Presidential address: Society of university surgeons, Recruitment and retention of academic surgeons, *Surgery* **86**(1):11.

14

Clinical Trials of New Drugs

Special Problems

MARTIN ROGINSKY

Drug Termination

Terminating an investigational drug in a clinical trial carries with it certain risks for subjects that are too often disregarded in the research protocol despite clinical indications that this critical posttrial period be anticipated prior to initiating the study. The potential for adverse consequences, as well as ethical considerations underlying all human experimentation, make it essential to consider special areas of concern relevant to the postadministration phase of the clinical trial. There are specific ethical and moral rights of subjects which may be jeopardized at trial conclusion, and there are direct harmful effects that may occur only in the posttrial period. These two concepts may be further divided into the following separate topics that warrant individual discussion:

1. The failure to include the posttrial period in the requirements of informed consent.
2. The failure to recognize that psychological and physical difficulties may result from the emotional consequences of loss of the trial agent.
3. The failure to consider potentially negative and harmful physical effects subsequent to agent withdrawal.

MARTIN ROGINSKY ● Division of Endocrinology and Metabolism, Nassau County Medical Center, East Meadow, New York 11801, and Department of Medicine, State University of New York at Stony Brook, Stony Brook, New York 11794.

4. The failure to take account of long-term effects of an agent's administration that are neither predictable nor foreseeable from short-term trials.

Although these topics will be examined separately, they are closely interrelated as elements of a singular concern that must prevail for the safety and well-being of the individual, healthy or not healthy, paid or unpaid, who volunteers as a human subject in a clinical trial (Trout, 1976).

Despite the well-publicized adverse incidents that have resulted from termination of a therapeutic trial, the parties most immediately concerned have not effectively dealt with the problems. While the investigator, the Food and Drug Administration (FDA), the sponsor, and the IRB must all share responsibility for consequences of not having adequately prepared for potentially adverse effects (Sadusk, 1974), it is the biomedical investigator and the IRB that presently must provide the primary impetus to mitigate such difficulties. Although these problems could be attenuated by permanent solutions in the future, there are essential first steps that can be taken now, and these are discussed below.

Inclusion of the Posttrial Period within Informed Consent

The first area of specific concern is assurance that the informed consent clearly defines the boundaries of the trial period and specifically includes the posttrial experiences. Requirements on informing subjects about the posttrial period are singularly neglected by current regulations guiding both the IRB and the investigator. Although there is no universal agreement on what constitutes sufficient understanding for a truly informed consent (Lebacqz and Levine, 1977), Title 21 of the Code of Federal Regulations (CFR) section 130.37(h) stipulates the following:

> The investigator should make known to the subject the nature, expected duration, and purpose of the administration of the investigational drug, the method and means by which it is to be administered, the hazards involved, and existence of alternative forms of therapy, and the beneficial effects upon his health or person that may possibly come from the administration of the investigational drug.

There is no requirement* that the prospective subject be informed that upon conclusion of the trial the therapeutic agent under investigation will/may

---

*Effective July 27, 1981, the FDA has added an additional element of informed consent that directs the investigator when appropriate to inform each subject "of anticipated circumstances under which the subject's participation may be terminated by the investigator without regard to the subject's consent." This alternative requirement, however, falls short of advising the subject of the possible harmful consequences that may result upon termination of a trial agent. (*Federal Register*, January 27, 1981, p. 8951.)

"It may very well bring about immortality,
but it will take forever to test it."

Reprinted with permission from *Research Resources Reporter* Volume 5, May, 1981 (© Sidney Harris, 1980).

be withdrawn, that this withdrawal will be independent of any efficacy or improvement in the patient's condition while receiving such therapy, and that he or she may be denied access to this agent or technique until such time as it becomes commercially available. Unfortunately, the time period between testing and commercial availability is highly unpredictable, but it usually takes years until the drug is approved for marketing by the FDA, if it is approved at all (Sadusk, 1974; Wardell, 1974; Lasagna and Wardell, 1975). Therefore, it is essential that in the process of properly informing the patient/subject the investigator and IRB address the posttrial period. Although this additional information may cause some prospective subjects to decline participation, this information will both better prepare the patient for withdrawal of the investigational agent and reduce possible hostile confrontations at the conclusion of the trial.

Consequences of Drug Withdrawal

Emotional Consequences

The second area of concern is the possibility that the patient may experience an emotional reaction in response to the stress evoked by termination of the trial drug. This can occur despite efforts to properly inform the subject of this eventuality (editorial: *Psychosom. Med.*, 1976). The potential for negative emotional reactions at a trial's conclusion may be caused by a variety of factors, such as the subjects' personality and the attitudes of the investigator and staff and their relationship with the subjects. Adverse emotional reactions, particularly if there is the possibility of untoward consequences following termination of the trial, must not be ignored. These psychological and psychosomatic reactions to withdrawal of an apparently successful investigational agent can ultimately be counterproductive to an entire study.

Perhaps cardiovascular disease, and specifically clinical evaluation of the antianginal and antihypertensive agents, best illustrates this potentially pernicious phenomenon. The role of the psyche in the initiation, prolongation, aggravation, and complication of cardiovascular diseases, where these drugs are effective, has been studied quite extensively. Despite wide acceptance of the critical role emotional and behavioral stress plays in coronary disease and hypertension, the FDA's 1977 guidelines for clinical evaluation of antianginal drugs—the purpose of which was to present current approaches to human subjects research—failed to mention the potential for adverse emotional reactions upon termination of the trial.

Negative Medical Consequences: The "Rebound Phenomenon"

The third area of specific concern, and one of more direct and immediate medical import, relates to the risk of intensification of disease or even the development of new manifestations resulting, directly or indirectly, from the precipitous withdrawal of a therapeutic agent (Alderman, 1974; Webster, 1974; Burden and Alexander, 1976; Strauss, 1977; Shand and Wood, 1978). These have been noted with reference to a variety of prescribable drugs, all of which have had extensive clinical trials. This "rebound phenomenon" is increasingly being reported in the literature. Webster (1974), in reporting his observations of "rebound hypertension" associated with the withdrawal of one FDA-approved antihypertensive agent, noted that this occurred regardless whether the drug had been terminated abruptly or the dosages gradually reduced over a period of time. Furthermore, in either case, the phenomenon was shown to resist readministration of the withdrawn agent. In contrast to expectations based on the natural history of conditions such as angina or hypertension, the emergence of symptoms more severe than those of pretreatment observations could not be simply the result of a deteriorating condition whose symptoms had been suppressed during treatment with such drugs.

Lasagna (1975) and Mundy (1974) have recommended that drug usage studies concerning adverse reactions to drug withdrawal be conducted as soon as the drug is marketed. It is suggested here, however, that such studies be conducted with participating subjects consequent to the "conclusion" of the formal "Phase III" trial, and during the time necessary for the sponsor to prepare and the FDA to evaluate the new drug application. This would upgrade the level of information concerning general public safety from insufficiently tested drugs, and would ensure continued treatment with the test drug if the physician/investigator determines it to be the treatment of choice.

The investigator may find himself in an ethically compromising situation if, upon completion of the trial, the data indicate that the test drug is the treatment of choice for some or all who participated as subjects. An investigator operating under FDA regulations, must withdraw the drug and is shielded from liability for the consequences of such withdrawal since he is acting according to the law. As a physician, however, he is ethically charged with the physical and psychological well-being of his patients. The role conflict is obvious, and Hodges (1974), in discussing the dual role of the physician/investigator in his relationship with the patient/subject, contends that if the subject is also a patient the priorities of physician–patient conduct must dominate. The medical literature contains many ethical objections to placebo-controlled drug trials on the basis that placebos entail the selective withholding of potentially therapeutic agents from patient/subjects in need (Hill, 1963; Hodges, 1974; Peck, 1975; Lorber, 1975a,b). We suggest that the logic of these objections may be equally applicable to the practice of denying patient/subjects access to a test drug that they have been receiving for relatively long periods of time, and whose efficacy may be superior to available remedies.*

Long-Term Side Effects

The final concern is the failure to consider the development of late effects of an agent that are neither predictable nor foreseeable from a short-term trial. Currently, the FDA requires only a 30-day followup subsequent to discontinuation of a Phase III trial for most investigative agents, despite the fact that long-term effects of drug withdrawal must be individualized, particularly since no single condition and no single patient reaction to a specific drug can always

---

*Final FDA regulations state, in its basic principles of ethical conduct for clinical research, that "in any medical study, every patient—including those of a control group, if any—should be assured of the best proven diagnostic and therapeutic methods." (*Federal Register,* January 27, 1981, p. 8953.) This contradicts the FDA's general recommendation for placebo control in clinical studies, and seems to indicate a more cautiously observed posttrial period with the allowance for continuation of the trial agent where indicated.

be predicted. There are many examples of untoward consequences with therapeutic agents that were not recognized during the clinical trials. The recent experience (FDA Drug Bulletin, 1980; Letters to the editor: *NEJM,* 1979) with ticrynafen (Salacryn®), a uricosuric diuretic, clearly illustrates a potentially adverse outcome from adherence to this policy. The serious nephrotoxicity related to this drug received scrutiny only following wide commercial distribution. The drug was subsequently withdrawn from clinical practice.

Within the context of these problems, merely informing prospective subjects that they are to be treated with an experimental agent which will be eventually withdrawn falls far short of the spirit of informed consent, the right of the patient to self-determination, and the exercise of free will (Schiffrin, 1977). Including information on termination of the drug in the consent procedure may or may not diminish adverse emotional reactions. It would, however, have no effect on possible delayed adverse responses to the drug during the trial or beyond the trial period.

The ultimate resolution of problems associated with terminating investigational drug trials will require serious conceptual and systemic changes involving all participants in biomedical research. Currently, the main obstacles to a resolution are (i) regulatory restrictions that delay the commercial availability of experimental drugs, despite recent FDA interest in expediting accessability of important new drugs (Finkel, 1980); and (ii) problems of economic and practical plausibility, i.e., the cost of monitoring continued administration of an experimental drug compounded by the cost and physical limitations on medical services and facilities essential to its distribution. While these problems are not likely to be resolved in the very near future, it is obvious that responsibility for a subject's condition is not limited to the time period during which the subject may be participating in the trial.

Despite the hazards posed for subjects during the posttrial period, those who are ultimately responsible for the ethical conduct of clinical research have failed to provide adequate structure for observing and controlling the long- or short-term effects of drug termination. This neglect on the part of the regulatory agencies is particularly puzzling in view of undue complications of test drug withdrawal that recently occurred. With the actions of new drugs becoming more complicated, the unknowns associated with their use will become increasingly precarious and will, undoubtedly, magnify risks to subjects.

Use of Placebos

Employment of a placebo during a clinical trial presents both the physician/investigator and the IRB with an ethical and moral dilemma. This dilemma, inherent in many double-blind studies, is based on the procedure whereby selected subjects, who are generally patients in need of effective medical treatment, are "treated" with either an inert substance, a mock procedure, or some otherwise less potent treatment for the purpose of comparison to the trial agent. While this approach is well-recognized and accepted for the collection of valid data, and it is considered by some to be a prerequisite of the scientific method, it may deprive the patient of an equal opportunity for effective therapy. Clinical investigators and members of an IRB, for whom the dictum *primum non nocere* is inviolate, may experience conflict with the goals of biomedical investigation, the biometric and biostatistic objectives of regulatory agencies, the assurances needed by commercial sponsors, and the models for the conduct of research promulgated by the scientific community (Hodges, 1974). Since the use of placebo control is considered virtually indispensable in clinical research, it is necessary to identify conflicting responsibilities, and to understand the historical origins of placebos, the factors which have led to their wide use in clinical trials, and the ramifications of the controversy surrounding their use. The following brief summary of these issues includes references to several excellent review articles to which the reader is referred for a more extensive examination of the issue.

The "Placebo Effect"

Historically, the *placebo* (from the Latin verb, "I shall please") was traditionally dispensed as an imitation or inert medication to placate patients for whom no underlying pathology could be detected as the cause of their complaints (Brody, 1980). The concept that symptoms could be dismissed as imaginary if they were responsive to placebo therapy is still evident in current medical attitudes (Goldberg *et al.,* 1979). However, despite this tenacious characterization of the placebo as a "dummy" treatment, the *placebo effect,* defined as the subjective and objective responses of a patient to a "medical" treatment that lacks any direct or indirect bioactive properties, is a phenomenon which was not readily distinguishable from true medicinal results until scientific methods were applied under controlled conditions (Beecher, 1955; Wolf, 1959; Shapiro, 1968: Benson and McCallie, 1979). The placebo effect has highlighted a number of questions about the nature of illness and the efficacy of drugs versus other means of treating disease that have only been partially answered to date. The significance of the placebo and the placebo effect cannot be overemphasized. Shapiro (1968) has gone so far as to state that without the placebo as an instrument in biomedical research, we would have little understanding that the history of medicine itself has been the history of the placebo effect.

Continued experiences have led to the acceptance of the placebo effect as an integral aspect of all therapeutic interactions (Beecher, 1955; Cousins, 1979). In each interaction, there are three elements influencing the response to medical treatment: (i) the physician's expectations, communicated to the patient directly or indirectly, (ii) the expectations of the patient, and (iii) the entire therapeutic milieu that includes all aspects of the physician/patient relationship (Shapiro, 1964; Benson and McCallie, 1979). Acceptance of the placebo effect has prompted a need to discriminate between the subjective dynamics of the therapeutic situation and the objective effects of the therapy. The placebo has progressed from a form of medical deception to a phenomenon which involves the hopes and expectations of both subject and investigator engendered by the whole of the treatment process.

While the placebo effect is appreciated as a component of the total healing process, it has also been recognized as a potential source of bias. Unless the placebo effect is anticipated in the experimental design, seriously distorted conclusions concerning the true effectiveness of a new drug or therapy could result (Brody, 1977; Beecher, 1955; Wolf, 1959; Shapiro, 1964; Shapiro, 1968; Benson and McCallie, 1979; Dollery, 1979; Feinstein, 1980). Studies of the placebo effect over the past 30 years have revealed a consistent and reliable baseline effectiveness of approximately $35.2 \pm 2.2\%$ (mean $\pm$ S.D.) in tests dealing with a wide spectrum of physical and psychological disorders (Beecher, 1955; Benson and McCallie, 1979; Levine *et al.*, 1978). It has not been uncommon, in individual studies, to observe a placebo effect as high as 70–90%, an occurrence that would obviously confound statistical isolation of this phenomenon (Benson and McCallie, 1979; Beecher, 1955). Recognizing the extent of this effect, investigators have sought to identify and isolate "placebo responders," i.e., patients postulated as a stable proportion with identifiable underlying psychological features making them specifically prone to the effects of suggestion. Studies conducted along this line, however, have neither revealed consistent patterns that could assure identification and isolation of such patients, nor have they demonstrated any reliable means to predict a subject's response to placebo in a succession of applications (McNair *et al.*, 1979). In-depth analysis in this area has revealed not only positive benefits, but also adverse and toxic side effects with objective pharmacologic, physiologic, and/or anatomical consequences attributable to placebo administration alone, all of which further obscures objective evaluation of uncontrolled clinical trials (Brody, 1977; Beecher, 1955; Wolf, 1959).

Placebo Control

The need for placebo control in the correction of biomedical data can have additional moral and ethical implications. The conflict apparent in the physician investigator's relationship with the patient/subject may demand the accommodation of divergent goals (Hodges, 1974) and has been discussed elsewhere in this book. It is commonly accepted that the physician's most immediate responsibility must be directed toward the best interests of his patient. We also recognize that there is a more global obligation to discover new and better treatment for all patients suffering from disorders for which there is currently little or no relief (Hodges, 1974). Superficially, one might characterize the use of placebo control as unethical because of its obvious denial of apparent effective therapy, yet the calculation of the risks and benefits of the trial agent versus placebo may be obscure and complex. For example, the trial agent may expose the patient to a risk greater than those presented by the disease itself or may offer therapy that is less effective than another treatment already available. If a placebo control group is required as a means of assuring the validity of data, ethical considerations would suggest that the trial be designed for the possible benefit of patients who has availed themselves of all conventional therapy without success, or for whom there are no other reasonable alternatives with a greater degree of putative effectiveness.

The solution to this dilemma lies in some yet-to-be devised means of evaluating new treatments that would account for the placebo effect without having to deprive patients of potentially effective treatment. Advances in statistical and technical methodology along with a more thorough understanding of the dynamics and physiology of the placebo effect itself, may ultimately lead to an answer. Temporary solutions have been offered in isolated instances, but a global solution does not yet exist (Zelen, 1979; Healy, 1978).

Factors to Be Considered in Reviewing Placebo Research

In the absence of an ideal solution, both the investigator and the IRB must weigh a given protocol by current standards of biomedical investigative technique. It must be borne in mind that placebo therapy is not analogous to no treatment at all and that a reasonably high rate of positive response and/or symptomatic improvement may be expected from the use of a placebo. An IRB might be guided by the following questions when evaluating a new drug protocol involving a placebo control:

(1) How necessary is the proposed research in terms of what is to be gained in useful medical knowledge in comparison to potential risks presented to the subject? This is both a subjective and objective question and requires analysis of the usefulness of the knowledge to be gained and comparison with the potential for harm to the patient according to the severity of the illness, its stage of advancement, and the available alternatives.

(2) Is a placebo necessary in this situation? The IRB must assess whether or not a placebo control is actually required. For example, the comparison of an oral placebo with a surgical procedure is probably a

faulty design that will yield meaningless data. A trial that produces meaningless data wasting the subject's participation and his sacrifices is morally and ethically reprehensible. In many instances, the efficacy of a drug may be strictly dependent upon the proper adjustment of dosage according to a number of variables for each patient. In such cases, the fixed dosage requirement which is usually employed in double-blind studies may depreciate pharmacologic effects that would otherwise be detectable. Good study design will correct this failing. Little attention has been directed toward the drawbacks involved in placebo controls and with the increasing requirement by the FDA for the inclusion of placebo controls in evaluation of all new drugs. It is obvious that increasingly sophisticated protocols for placebo control will be forthcoming. One must avoid the trap of invoking a placebo control in order to avoid conflict with a regulatory agency as opposed to generating valid scientific data. The IRB is not required to "rubber stamp" an unreasonable demand made for the wrong reason. There also exists the possibility that a comparable drug already on the market can be used as a control in lieu of a placebo from which the patient could theoretically benefit. A test of the new agent versus this accepted therapy might yield the same data without raising the issue of placebo control.

(3) What jeopardy is portended for the patient/subject by his participation in the trial should he not receive the active agent? It must be remembered that the side effects of the active agent are a benefit to those patients who are assigned to the placebo group. In a crossover design, this consideration is probably irrelevant.

(4) Can the placebo control be kept small without compromising the integrity of the data?

(5) How will the investigator deal with subjects whose condition seriously deteriorates as a result of participating in a placebo control group? The FDA has suggested that placebo responders be isolated from the trial by screening methods beforehand in order to reduce the number of patients who might have otherwise benefited from the placebo effect. One might also allow provision for ongoing third party unblinded review in order to drop recipients of placebo therapy if excessive risk becomes evident.

Consideration of these questions may not eliminate the dilemma but may help to avoid some serious breaches of ethics involved in the experimental use of placebos that have occurred in the past. The ultimate safeguard against an abuse rests in the hands of the biomedical investigator who accepts or rejects a protocol, and in the hands of the IRB which makes a final review of the investigative decision. Although the placebo is an important and often necessary element of biomedical experimentation, it may be a danger when it is overexploited or applied indiscriminately. One cannot advocate abandoning the placebo even in light of the ethical conflict created for the physician/investigator. Starting from the viewpoint of the physician's primary obligation to heal, one must remember that participation in a clinical trial may elicit a powerful placebo effect for many patients, and that the healing process involves much more than a specific chemical compound. The use of placebo controls has enlightened us as to the holistic nature of medical treatment and made possible discrimination between many necessary components of health and healing.

References

Alderman, E. L., 1974, Coronary artery syndromes after sudden propranolol withdrawal, *Ann. Intern. Med.* **81**:625–627.

Beecher, H. K., 1955, The powerful placebo, *JAMA* **5**(159):1602–1606.

Benson, H., and McCallie, D. T., 1979, Agina pectoris and the placebo effect, *N. Eng. J. Med.* **300**:1424–1429.

Brody, H., 1980, *Placebos and the Philosophy of Medicine: Clinical, Conceptual, and Ethical Issues,* University of Chicago Press, Chicago.

Burden, A. C., and Alexander, C. P. T., 1976, Rebound hypertension after acute methydopa withdrawal, *Br. Med. J.* **1**:1056–1057.

Cousins, N., 1981, *Anatomy of an Illness as Perceived by the Patient,* p. 176, Bantam, New York.

Editorial, 1976 The heterogeneity of "psychosomatic" disease, *Psychosom. Med.* **38**:371–372.

Finkel, M. J., 1974, Investigational drug studies: Recent FDA efforts, in: *Principles and Techniques of Human Research and Therapeutics* (G. F. McMahon, ed.), pp. 45–49, Futura, New York.

Finkel, M. J., 1980, The FDA's classification system for new drugs: An evaluation of therapeutic gain, *N. Engl. J. Med.* **302**(3):181–183.

FDA Drug Bulletin, 1980, February 10, Number 1, pp. 3–4. Goldberg, H. L., and Finnerty, R. J., 1979, Comparative efficacy of torfisopan and placebo, *Am. J. Psychiatry* **136**:196–199.

Healy, M. J. R., 1978, New methodology in clinical trials, *Biometrics* **34**:709–710.

Hill, A. B., 1963, Medical ethics and controlled trials, *Br. Med. J.* **1**:1043–1049.

Hodges, R. M., 1974, Ethical considerations in clinical research, in: Principles and Techniques of Human Research and Therapeutics (G. F. McMahon, ed.), pp. 31–38, Futura,

Lasagna, L., 1975, Clinical trials of drugs from viewpoint of the academic investigator (a satire), *Clin. Pharmacol. Ther.* **18**:629–633.

Lasagna, L., and Wardell, W. M., 1975, Commentary: The FDA, politics, and the public, *JAMA* **232**:141–142.

Lebacqz, K., and Levine, R. J., 1977, Respect for persons and informed consent to participate in research, *Clin. Res.* **25**:101–107.

Letters to the editor, 1979, *N. Engl. J. Med.* **301**:1179–1181.

Levine, J. D., Gordon, N. C., and Fields, H. L., 1978, The mechanism of placebo analgesia, *Lancet* **2**:654–657.

Lorber, M., 1975a, When is double-blind evaluation improper? *J. Clin. Pharmacol.* **15**:84.

Lorber, M., 1975b, Delaying double-blind drug evaluation in usually fatal diseases (Pelter), *N. Engl. J. Med.* **293**:508–509.

Mundy, G. R., 1974, Current medical practice and the Food and Drug Administration: Some evidence for the existing gap, *JAMA* **229**:1744–1748.

McNair, D. M., Gardos, G., Haskell, D. S., and Fischer, S., 1979, Placebo response, placebo effect, and two attributes, *Phycopharmacology,* **63**:245–250.

Peck, A. W., 1975, Letter and ethics of randomized trials, *N. Engl. J. Med.* **293**:1270.

Sadusk, J. F., Jr., 1974, The effect of drug regulation on the development of new drugs, in: Principles and Techniques of Human Research and Therapeutics (G. F. McMahon, ed.), pp. 58–68, Futura, Mt. Kisco, N.Y.

Schiffrin, M. J., 1977, The regulation of clinical research, *J. Clin. Pharmacol.* **17**:686–690.

Shand, D. G., and Wood, A. J. J., 1978, Propranolol withdrawal syndrome—Why? (editorial), *Circulation* **58**(2):202–203.

Shapiro, A. K., 1964, Etiological factors in placebo effect, *JAMA* **187**:712–714.

Shapiro, A. K., 1968, Study of the placebo effect with a placebo test, *Comp. Psychiat.* **9**:118–137.

Strauss, F. G., 1977, Withdrawal of antihypertensive therapy. Hypertensive crisis in renovascular hypertension, *JAMA* **238**:1734–1736.

Trout, M. E., 1976, Ethics of clinical research, *Conn. Med.* **40**(4):201–204.

United States Department of Health, Education, and Welfare, 1977, Guidelines for the clinical evaluation of anti-anginal drugs, HEW(FDA) 78–3047.

Wardell, W. M., 1974, Drug development, regulation and the practice of medicine, *JAMA* **229**:1457–1461.

Webster, J., 1974, Withdrawal of antihypertensive therapy, *Lancet* **2**:1381–1382.

Wolf, S., 1959, Placebos, *Res. Pub. Assoc. Res. Nerv. Ment. Dis.* **37**:147–161.

Zelen, M., 1979, New design for randomized clinical trials, *N. Eng. J. Med.* **300**:1242–1245.

Psychiatric Research

John M. Kane, Lewis L. Robbins, and Barbara Stanley

The task of the IRB in reviewing research protocols involving patients with psychiatric illness is among its most difficult responsibilities. Ethical concerns applicable to other areas of medical research—concerns about the physician-patient relationship, the effect of illness on one's mental state, the detrimental effect of long term hospitalization on one's sense of autonomy—all seem to become heightened in the mind of the lay public as well as many professionals when the research involves psychiatric patients. The very nature of psychiatric illness calls into question the competence of these potential subjects to give their consent. In addition, the kind of information often essential to many psychiatric research studies can be highly personal and may require special consideration in order to respect the privacy of subjects and to maintain strict confidentiality. Therefore, psychiatric research protocols may require additional scrutiny by the review board even for "no-risk" research projects.

This chapter focuses on the special issues the IRB must consider as it evaluates research projects involving psychiatric patients. These issues include the questions concerning the competence of psychiatric patients to give informed consent; the problems of institutionalization of psychiatric patients; the problem of voluntariness; the issue of incomplete disclosure of information to the patient; confidentiality of patient data; and alternative models of consent (i.e., third party observers, proxy consent). Although this chapter discusses issues particular to the psychiatric population, it must be remembered that this

JOHN M. KANE AND LEWIS L. ROBBINS • Department of Psychiatry, Long Island Jewish–Hillside Medical Center, New Hyde Park, New York 11042, and Department of Psychiatry, State University of New York at Stony Brook, Stony Brook, New York 11794. BARBARA STANLEY • Department of Psychiatry, Wayne State University School of Medicine, Lafayette Clinic, Detroit, Michigan.

population has much in common with other patient populations. Therefore issues such as risk/benefit ratio of the proposed research must be addressed as with nonpsychiatric research.

Tests of Competency

It has been frequently assumed that psychiatric patients, like children, need special protection because they may lack the capacity for making mature and informed judgments. It has also been assumed that they are easily influenced by researchers' presumed conflicts of interest. These concerns have led to controversial suggestions; among them is the proposal that impartial auditors and advocates be involved in various stages of psychiatric research, particularly in securing the subject's consent (*Federal Register,* 1978). Clearly, such extreme positions may do as much to deprive psychiatric patients of their autonomy as well as important new knowledge and treatments as they do to protect them from potential risks (Eisenberg, 1977). Those patients who are presumed to be least able to give informed consent (e.g., chronic schizophrenics who are unresponsive to available treatment, individuals with senile dementia, mental retardation or extreme behavioral disturbances) are, in fact, those most in need of new knowledge and treatment resulting from research.

As a result of this potential difficulty, it is important that IRBs assess the clarity of the informed consent and the probable competence of the subject population. With respect to this latter issue, competency, the assumption has frequently been made that psychiatric patients as a group may not be "competent" to give informed consent. Further, many would assume that there is a positive correlation between level of psychopathology and inability to give informed consent. In other words, it is a popular notion that patients manifesting greater psychopathology are less able to give consent. The validity of these assumptions will be discussed later in the chapter. In order to discuss the issues surrounding competency, we must first describe the tests commonly used for assessing it.

Roth *et al.* (1977) have reviewed the following five tests of competency:

1. "Evidencing a choice" suggests that "the competent patient is one who evidences a preference for or against treatment. This test focuses not only on the quality of the patient's decision but on the presence or absence of decision."
2. "Reasonable" outcome of choice assesses the patient's capacity to reach a reasonable or responsible decision. This test emphasizes the "reasonableness" of a decision rather than the process involved in making the decision.
3. Choice based on "rational" reasons suggests that the actual decision-

making process of a patient be examined in order to determine its quality or logical nature. It has been pointed out that this test is problematic in assuming that the actual decision made is a result of the reasons the patient gives. Further, although a patient may make a choice that is reasonable, his or her stated reasons may be "irrational." Thus, the results of this competency test may conflict with the "reasonable outcome" test. Another difficulty with this test has been suggested by Roth *et al.* (1977); the emphasis on rational reasoning can too easily become a global indictment of the competency of mentally disordered individuals justifying widespread substitute decision-making for this group.

4. The "ability to understand" criteria suggests that the patient manifests an adequate ability to understand the issues involved, the information provided, and so forth, without necessarily weighing them in a "rational" manner or reaching a "rational" decision. Level of understanding, in this sense, can be tested in a somewhat objective fashion, but what level of understanding is sufficient remains at issue.

5. "Actual understanding" implies a specific test of whether or not the patient has actually understood the issues involved in the decision at hand. These last two competency tests which focus on understanding are very closely aligned.

Competency of Psychiatric Patients

There is a paucity of empirical data examining the competency of psychiatric patients as a group. This lack of empirical research is surprising considering that there are proposals that would offer special protection for the institutionalized mentally disabled as a group (National Commission, 1978). Some research studies have been conducted on this topic but they are not comparative in nature and therefore do not have a direct bearing on this question (Grossman and Summer, 1980; Roth *et al.*, 1980). In order to address this issue, a comparative study was undertaken to empirically assess the capacity of the mentally ill to engage in the consent process as contrasted with other patients (Stanley *et al.*, 1981). This study will be described in some detail because of the limited empirical data available. The research investigated the following questions: (i) Do the mentally ill expose themselves to greater research risks than nonpsychiatric patients? (ii) Do they refuse to participate in projects with highly favorable risk/benefit ratios more often than patients? (iii) Does degree of psychopathology, a frequently used indicator of capacity to give consent, relate to willingness to participate in overly risky research?

Twenty-seven psychiatric patients on a locked unit at a psychiatric hospital and thirty-eight medical patients on a general medicine unit participated in this study. The psychiatric patients manifested severe psychopathology—primarily with admitting diagnoses of schizophrenia and psychotic depression. The diagnoses of the medical patients were typical of those on general medicine units including various heart, pulmonary, or liver ailments. Patients manifesting organic brain syndrome or mental retardation were excluded from the study in order to avoid a potential source of ambiguity.

To assess willingness to be exposed to research risks, a series of hypothetical research studies were presented to the patients. Following each research description, patients were asked a series of questions assessing their willingness to participate in each study. Following this interview, the patients' psychopathology was evaluated independently by a psychiatrist using a psychiatric rating scale. The vignettes varied in level of risk and benefit ranging from high risk/low benefit to minimal risk/high benefit. They were written in relatively simple language to ensure that the reading level was within the average individual's capacity and incorporated elements of a standard consent form: purpose, procedure, risks, and benefits.

Figure 1 shows the percentage of patients in both groups combined willing to participate in each study. The results show that agreement varied in a way consistent with the amount of risk as previously judged by a health professional and student sample. Studies A and D, low risk projects, elicited the highest percentage of participation. These were followed by the moderate risk study (Study F) and high risk studies (Studies B, C, and E), respectively.

The percentages of psychiatric and medical patients agreeing to each

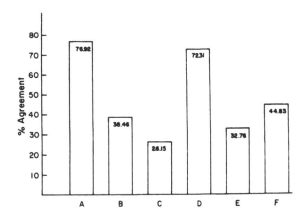

Figure 1. Agreement to participate in six hypothetical studies by total sample. A, Low risk/high benefit; B and C, high risk/low benefit; D, low risk/high benefit; E, high risk/high benefit; F, moderate risk/high benefit. From Stanley *et al.* (1981), reprinted by permission, copyright of The American Psychiatric Association.

study were not significantly different (Fig. 2). This seems to indicate that psychiatric and medical patients evaluate research risks and benefits in a similar manner. Further, overall willingness to participate in research projects was examined by summing the number of studies each patient agreed to participate in. No relationship was found between type of patient (medical or psychiatric) and number of studies a person agreed to.

Willingness to participate in high risk versus low risk studies was also examined (Fig. 3). Again, no significant differences between groups were found. Psychiatric patients neither agreed to participate in high risk studies more often nor did they more frequently refuse studies with highly favorable risk/benefit ratios. With respect to agreement to participate in these studies the psychiatric patients behaved similarly to the medical patients.

Degree of psychopathology as a useful indicator of ability to evaluate risks was also examined. No relationship was found between level of psychopathology and number of studies agreed to for the psychiatric patients. In addition, when looking only at high risk/low benefit studies, no relationship was found between the patients' psychopathology and agreement to participate for psychiatric patients. The results of this study indicate that level of psychopathology may not be a good index of patients' ability to consent to research. The findings in this study contradict the frequently made assumption that the more severely disturbed the patient is, the less able he is to evaluate research risks. If further research confirms these findings, it may call into question the view that psychiatric patients are inherently imcompetent.

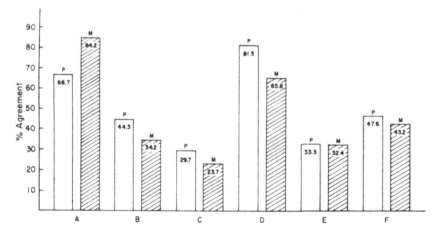

Figure 2. Psychiatric and medical patient agreement to participate in six hypothetical studies. Column key as for Fig. 1; P, psychiatric inpatients; M, medical inpatients. From Stanley *et al.* (1981), reprinted by permission, copyright of The American Psychiatric Association.

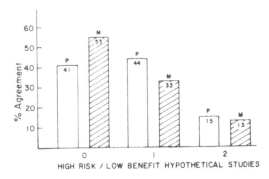

HIGH RISK / LOW BENEFIT HYPOTHETICAL STUDIES

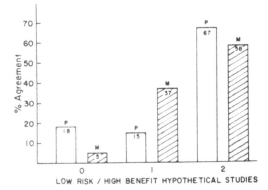

LOW RISK / HIGH BENEFIT HYPOTHETICAL STUDIES

Figure 3. Participation in hypothetical studies: High risk/low benefit and low risk/high benefit. From Stanley *et al.* (1981), reprinted by permission, copyright of The American Psychiatric Association.

The Problem of Voluntariness

A continued source of confusion among those concerned with these problems, including the courts, is the delineation of competency and voluntariness. The latter implies that no coercion or improper inducements are involved in subject consent. This concept becomes quite complicated when one considers the spectrum of voluntariness, which ranges from excessive coercion to covert patient wishes to please the doctor. As an example, it is not unheard of for a patient to state that he feels he will get more clinical attention if he participates in a research project. Does this represent favored treatment, or is it a recognition of fact that monitoring for drug side effects and other related procedures can be more frequent and more detailed in an "experimental" setting than in standard clinical practice? Perceptions of better care may induce nonpsychiatric patients to participate in research as well. The IRB must therefore ensure that there is no marked difference in the setting that research subjects enjoy as compared to other patients who are not research participants. Equivalent sleeping quarters, food, medical care, clothing, recreational activities, and other services should be provided for all patients regardless of their decision to join a research project. If this is not the case, the setting itself could prove coercive.

Given recent litigation involving the right to refuse treatment, the question arises as to whether patients who are involuntarily hospitalized are competent to give consent. The question of voluntarily entering an institution or being involuntarily committed is discussed by Annas *et al.* (1977) from the dual viewpoints of providing standard medical treatment and research. They point out that "the entire distinction between voluntary and involuntary hospitalization is often murky. The majority of voluntary admittees enter 'voluntarily' only under the threat of involuntary commitment.... In some cases, ill individuals who may be mentally capable of managing for themselves are institutionalized because family and public agencies are unwilling or unable to assist them in finding suitable living arrangements. Such patients, who may be technically involuntarily committed, should be free to decide whether or not to join an experimental research project or accept an experimental therapy.

Alternative Models of Consent

Proxy Consent

While a patient's confinement status may not necessarily determine his or her competency to participate in a consent process for research involvement, there is increasing legal pressure to obtain proxy consent. However, it is not clear how proxy consent should be obtained. Physicians who traditionally filled this role are no longer trusted (for a variety of reasons, one of which is presumed conflict of interest) and relatives, as well, may have conflicting interests about treatment and research. Therefore, it is conceivable that the courts will increasingly assume the proxy role with an adversary situation being fostered.

Even if guardianship or proxy consent is obtained, the IRB must determine what degree of risk and benefit justifies this approach. One would assume that such circumstances would necessitate great direct potential benefit to the individual. In addition, even though the consent process might then bypass the subject an explanation and discussion of the research should be carried out with the individual to the extent possible. The IRB must be aware of the complex considerations required to ensure voluntariness and should reject extreme remedies in this context.

The Use of Third Party Observer to the Consent Process

Review boards have frequently called upon professionals not associated with the research endeavor to determine whether or not a specific patient is "capable" of giving informed consent, yet what this means is far from clear.

In practice, this frequently means that the psychiatrist directly responsible for the patient's care and not involved in the research makes a subjective judgment as to whether or not the patient is competent to give his consent. It would seem that the role of the IRB is to ensure an acceptable risk/benefit ratio in any protocol being applied to psychiatric patients. This would appear to be the most important and meaningful safeguard against potential abuse of patients with diminished competency while more objective standards to determine are being formulated.

If one were to accept the need for a third party observer to assess whether or not the patient has understood the nature of the project and its attendant risks and benefits, a case could be made for such an observer in a variety of other medical situations as well. However, prior to the institution of third party observers a number of issues should be explored including the necessity of observers, their utility, the cost of such a system and the amount of additional protection they provide.

In the case of psychiatric patients it may be helpful to involve the patient's family in the consent process, if feasible. This can help a patient reach a thoughtful decision through family discussions. However, the review committee must be aware that at times family members can be coercive, and this situation should be avoided. This is a difficult situation because it is hard to distinguish between reasonable or justifiable persuasion and inordinate influence or coercion.

Confidentiality

Confidentiality is an assumed right that must be safeguarded, particularly among psychiatric patients. Loss of privacy, in this context, can result in a significant change in "social reputation." The IRB should be alert to the sensitivity of this issue in present day society. It is also important, however, to distinguish the risk incurred in the context of research from the risk to confidentiality inherent in treatment itself. Much information involving psychiatric history-taking and data collection is of an extremely personal nature. Many individuals would see this information as carrying with it considerable potential stigmatization and prejudicial treatment. A great deal of research is psychopathology and deviant behavior does not carry with it any risk to the individual subject unless a breach of confidentiality occurs as a result of the research. It has been suggested (Robbins, 1977) "that privacy regulations which were not designed with research in mind can impede research considerably." Frequently researchers are interested in individuals only as a representative of a class of persons rather than as an individual, whereas employers, creditors, and the like are interested in him as an individual. Robbins (1977) has suggested that those regulations which are intended to protect human subjects have reached the point where "the researchers very act of compliance with the regulations while doing follow-up studies is likely to do the subject more harm than are the research procedures themselves."

Consent Form/Consent Process

Many IRBs have become more concerned with the "form" of the consent as opposed to the content. While the form is important, the process between the investigator and patient should also be considered. In dealing with protocols involving psychiatric patients the content must be evaluated in terms of both potential positive and negative impact. Loftus and Fries (1979) have cautioned us on the "dangers" of consent forms in arousing unnecessary levels of anxiety and producing undesired adverse reactions, particularly in highly suggestible populations. This may lead to a premature conclusion that psychiatric patients may become even more upset than other patients because of their emotional problems. However, empirical research has found that these conclusions are not warranted. Detailed information given to patients may serve to reduce anxiety rather than heighten it (Denney *et al.*, 1979).

A major concern of the IRB should be that consent forms are understandable to the average person. A survey by Gray *et al.* (1978) found that many consent forms were written in academic or scientific language. Consent forms must be carefully reviewed by laymen to assure their contents are not too technical. As laymen continue to serve on IRBs their own sophistication and familiarity with medical terminology increases and even they may begin to accept overly technical language, since it has become familiar to them. The board must be aware of this tendency. As a remedy, readability tests can be applied to consent forms to determine their complexity (Flesch, 1978; Fry, 1968).

The survey of Gray *et al.* (1978) also found that many IRBs confine their attention to the actual form or piece of paper to be employed in documenting informed consent, rather than the overall process of how the informed consent is obtained. There are several elements stipulated as essential inclusions in consent forms according to DHHS regulations (e.g., the purpose of the research, the procedure involved, the risks and benefits, a statement that the subjects are free to withdraw from the research, and an open invitation to ask questions). However, the context in which the consent form is discussed and the responsiveness and sensitivity of the investigator will have a significant effect on whether these principles are meaningfully applied. It is clear that informed consent should be, in fact, a continuing process and not merely a signature on the dotted line. The process occurs in the context of a complex relationship between a doctor/investigator and patient/subject.

The Physician–Patient Relationship

For patients experiencing such psychological phenomena as severe mood shifts, impairment in concentration, and continuing alterations in relatedness and sense of trust, the ongoing relationship with the investigator is of enormous importance. This applies also to the psychological stresses and changes occurring in response to physical illness. While one frequently considers these psychological phenomena with psychiatric patients, one should ask the same questions about individuals having open heart surgery, recovering from heart attacks, or being treated for cancer. The impact of these stresses on the ability to participate fully in the consent process requires further study.

Jonas (1969) has emphasized the fact that physical and emotional state, dependency in relation to the doctor, and the submissive attitude involved in any treatment "makes the sick person inherently less of a sovereign person than the healthy one." As Stone (1979) suggests, "one does not have to be a psychiatrist to know that the vulnerability and helplessness of sick people manifest in their relation to their doctors will not often be resolved by information."

The whole concept of informed consent in medical research remains problematic. There are few guidelines as to what constitutes an informed prospective research subject. Should the individual be told (i) what a reasonable individual would want to know, (ii) what a reasonable physician thinks he should know, or (iii) what the average physician in a nonresearch situation would actually tell a patient? What level of understanding should the individual have of the information provided? To complicate judgments as to what information is essential, there is very little evidence as to what role "information" plays in the decision-making process, and some research suggests that disclosure of information plays a minor role (Fellner and Marshall, 1970).

Incomplete Disclosure

The Belmont Report (1978) acknowledges that "a special problem of consent arises when informing subjects of some pertinent aspect of the research is likely to impair the validity of the research." The report goes on to state that "in many cases, it is sufficient to indicate to subjects that they are being invited to participate in research of which some features will not be revealed until the research is concluded. In all cases of research involving incomplete disclosure, it seems reasonable that such research is justified only if it is clear that (1) incomplete disclosure is truly necessary to accomplish goals of the research; (2) there are no undisclosed risks to the subjects that are more than minimal; and (3) there is an adequate plan for debriefing subjects, when appropriate, and for dissemination of research results to them. Information about risks should never be withheld for the purpose of eliciting the cooperation of subjects, and truthful answers should also be given to direct questions about the research. Care should be taken to distinguish cases in which disclosure would destroy or invalidate the research from cases in which disclosure would simply inconvenience the investigator."

It is sometimes argued that in some cases "full disclosure" can prove untimely or unduly upsetting to an individual, and can lead to rejection of a form of treatment that may hold therapeutic promise. However, this is not yet proven and we have not identified which patients are most vulnerable to such effects.

The Role of the IRB

As Robert Merton (1967) has suggested, social institutions serve "latent" as well as "manifest" functions. Once a bureaucratic structure has been established it will be utilized in ways and for purposes not originally anticipated and for which it was not originally designed. It is clear that the establishment of IRBs has served to function as a protective device for human subjects, but can also be used to protect investigators and institutions. These purposes are not necessarily incompatible, but those involved in IRBs should be aware of what functions their work serves, in order to be alert to potential conflicts.

It is useful for investigators to serve on IRBs even if only for brief periods in order to understand as much as possible about the perspective from which the board operates. Too often investigators see the review board as being in an adversary role, and no doubt some review boards foster this impression. It is essential that the committee perform an educative function as well as a review function. This ultimately makes the task of all concerned much easier. It is helpful for the investigator/clinician to help educate members of the review board as well. Review board members vary in their experience and preparation to serve on such committees, and few formal training programs are available for prospective members. Given the importance of the function that these bodies serve, perhaps more attention to specialized training is warranted. The importance of the psychiatric perspective cannot be emphasized enough, as one area of risk which must always be considered is psychological risk, not just for psychiatric research but for all biomedical studies.

It is also clear that health care professionals should not abdicate their responsibility to lawyers, ethicists, philosophers, judges, legislatures, and regulatory agencies. The importance of the clinical perspective in dealing with the complex issues involved in human subjects research should not be minimized or ignored.

The composition of the review board is critically important in the review of psychiatric research. Since many members may have little experience with psychiatric patients and their illnesses, and since psychiatric illness itself evokes a variety of conscious and unconscious reactions on the part of all of us, it is crucial to have adequate representation of mental health professionals on an institutional board. Much of what takes place in IRB discussions involves the sharing of appropriate information necessary for the committee to make required judgments and recommendations. The board should be constituted as broadly as possible. For hospitals and academic medical centers, this should pose no problem. Such settings provide ready access to a wide range of health professionals, all of whom are concerned with the welfare of the patient—psychiatrists, physicians in other medical specialties, social workers, and psychologists. In addition, IRBs are also likely to include clergymen, representatives of patient groups, students, and independent representatives of the community. IRBs in state or private institutions that may be located far from academic settings must make a particular effort to broaden their membership beyond their employees and the single nonaffiliated community representative required by the regulations. It is too easy to become desensitized to particular conditions or issues by exposure to a single point of view or set of assumptions in an intellectually or physically isolated setting. The research and the rights of patients are best served by an open intellectual environment. Institutions which may find themselves unable to constitute a review board appropriate to address the issues in psychiatric research should consider delegating the review process to the nearest academic medical center.

Finally, any medical research endeavor will involve simultaneous considerations of the best interests of the individual, and the best interests of that population or group of patients which the individual represents. As Eisenberg (1977) has reminded us, there has frequently been confusion of what is usual and customary in medical practice with what is safe and useful. The necessity for controlled investigations does not simply apply to the introduction of new or experimental approaches to treatment but the ongoing assessment and reasssessment of what is customarily practiced or prescribed. The history of medicine is replete with examples of treatments presumed to be "safe and effective" by the standards of the day, until evidence to the contrary becomes available.

References

Annas, G., Glantz, L., and Katz, B., 1977, *Informed Consent to Human Experimentation: The Subject's Dilemma*, p. 162, Ballinger, Cambridge.
Denney, M., Williamson, D., and Penn, R., 1979, Informed Consent: Emotional responses of patients, *Postgrad. Med.* **60**:205–209.
Eisenberg, L., 1977, The social imperatives of medical research, *Science* **198**:1105–1110.
Federal Register, 1978, Protection of human subjects, Volume 43, No. 223, p. 53950.
Fellner, C. H., and Marchall, J. R., 1970, Kidney donors: The myth of informed consent, *Am. J. Psychiat.* **126**:1245.

Flesch, R., 1978, A new readability yardstick, *J. Appl. Psychol.* **32**:221–233.

Fry, E., 1968, A readability formula that saves time, *J. Reading* **11**:513–516, 575–578.

Gray, B. H., Cooke, R. A., and Tannenbaum, A. S., 1978, Research involving human subjects, *Science* **201**:1094–1101.

Grossman, L., and Summers, F., 1980, A study of the capacity of schizophrenic patients to give informed consent, *Hosp. Commun. Psychiat.* **31**:205–207.

Jonas, H., 1969, Philosophical reflections on experimenting with human subjects, *Daedalus* **98**:219–247.

Loftus, E. F., and Fries, J. F., 1979, Informed consent may be hazardous to health, *Science* **204**:12.

Merton, R. K., 1967, *On Theoretical Sociology*, Free Press, New York.

Robbins, L. N., 1977, Problems in follow-up studies, *Am. J. Psychiat.* **134**:904–907.

Roth, L. H., Meisel, A., and Lidz, C. W., 1977, Tests of competency to consent to treatment, *Am. J. Psychiat.* **134**:279–284.

Roth, L., Lidz, C., Soloff, P., and Kaufman, K., 1980, Competency to consent to and refuse ECT: Some empirical data, paper presented at the Annual Meeting of the American Psychiatry Association, 1980.

Stanley, B., Stanley, M., Lautin, A., Kane, J., and Schwartz, N., 1981, Preliminary findings on psychiatric patients as research participants: A population at risk? *Am. J. Psychiat.* **138**:669–671.

Stone, A. A., 1979, Informed consent: Special problems for psychiatry, *Hosp. Commun. Psychiat.* **30**:321–326.

The Belmont Report, 1978, *Ethical Principles and Guidelines for the Protection of Human Subjects of Research*, National Commission for the Protection of Human Subjects of Biomedical and Behavioral Research, DHEW Publication No. (OS) 78–0012.

The National Commission for the Protection of Human Subjects of Biomedical and Behavioral Research, 1978, *Research Involving Those Institutionalized Mentally Infirm*, DHEW Publication No. (OS) 78–0006.

Irbs and the Regulation of Social Science Research

LAWRENCE SUSSKIND AND LINDA VANDERGRIFT

Introduction

The use of IRBs as a mechanism for regulating federally-funded research followed from the Public Health Service rules of 1966. These were issued in response to the public outcry against certain rather disturbing findings with some medical experiments.[1] The social science community was somewhat surprised, however, when the federal government insisted some years later that IRBs had responsibility for overseeing social science research as well as biomedical research involving human subjects. Although the original rules had ostensibly applied to social science research, neither the Public Health Service nor any of the early IRBs interpreted these regulations literally (Gray, 1979; Seiler and Murtha, 1979).

In 1974, the Secretary of Health, Education, and Welfare (in anticipation of Congressional passage of the National Research Act) issued further regulations regarding the protection of human subjects. By that time, substantial differences had arisen within the Department of Health, Education, and Welfare (HEW) regarding the applicability of the regulations to nonbiomedical research. It was understood within HEW (and alluded to in the preamble to the regulations) that further discussion and negotiations would be needed to construct regulations appropriate to social science research.[2] Separate regulations, however, were never issued. Internal conflicts within HEW were too difficult to resolve. Ultimately, the entire matter was dropped and the 1974

LAWRENCE SUSSKIND AND LINDA VANDERGRIFT ● Department of Urban Studies and Planning, Massachusetts Institute of Technology, Cambridge, Massachusetts 02139.

regulations were presumed to apply to social science research (Seiler and Murtha, 1980).

The National Research Act mandated that each institution involved in biomedical and behavioral research involving human subjects establish an IRB. Additionally, the Act established the National Commission for the Protection of Human Subjects of Biomedical and Behavioral Research. The Commission was responsible for evaluating the performance of IRBs and recommending ways of improving the review process. No social scientists were appointed to the Commission. The social science community was therefore caught unawares when the Commission concluded that prior review of both funded and unfunded research was in order, and that all social science research should be subject to the same review procedures as biomedical research (Seiler and Murtha, 1980). The Commission's recommendation and subsequent regulations proposed by HEW (now HHS) elicited wide comment from the social science community, much of it strongly negative. Proponents held that the IRB review process is a reasonable peer review mechanism that can help to protect subjects of social science research against risks to their psyche or dignity. As Caplan (1979) commented: "Peer review with public input seems a reasonable way to preserve both the rights and autonomy of potential research subjects and the rights of researchers to conduct their investigations in a free and unfettered manner." Caplan held that government agencies have a responsibility to the subjects of research to protect their privacy and dignity. Peer review seems the most painless approach.

However, there were those who argued strenuously that government regulations, originally conceived to protect subjects from the very real risks inherent in biomedical research, were being extended inappropriately to social science research and their extension constituted a threat to academic freedom and a form of prior restraint on free speech.[3] de Sola Pool argued that the proposed HEW regulation violated the "millenium old concept of a university":

> A university is a loose collection of independent scholars, not a staff of directed employees. HEW proposes to hold funds for universities hostage to the actions of each university member. If universities are to be thus responsible for studies that occur on their premises, they must convert their students and faculty into disciplined agents instead of what they have always been: independent thinkers guided by individual curiosity. (de Sola Pool, 1979.)

Current HHS regulations assume that it is the responsibility of the federal government to protect the public interest and the public's right to know by setting research standards for HHS funded research. While the final regulations are aimed at protecting the rights of human subjects by requiring informed consent, they substantially reduce the scope of previous regulations by exempting broad categories of research that present little or no risk to subjects. Exempted from regulatory coverage are social, economic, and educational research in which the only involvement of subjects will be in survey and interviews, observation of public behavior, or the study of data, documents, records, or specimens. *Exemption from review, however, does not exempt a researcher from obtaining informed consent.*

Clearly, not all research conducted by social scientists is harmless to subjects. In the 1950s a University of Chicago study which involved bugging federal juries brought on Congressional hearings and legislation. Dr. Stanley Milgram's well-known experiments in obedience caused considerable controversy among social scientists. In a study of Latin America funded by the Defense Department, Project Camelot, anthropologists found that their data could be misused and those who cooperated with them found themselves in jeopardy. It is not unusual for social scientists to involve themselves in a wide variety of controversial and sensitive areas, such as research on illegal aliens, tax evasion, sexual preference, as well as behavioral modification programs for the self-referred alcoholic, drug abuser, child or wife abuser, kleptomaniac and chronic gambler.

This chapter addresses the IRBs ethical concerns in reviewing social science research. Their concerns center on the deception of subjects, exploitation of certain subject classes, confidentiality and privacy, and avoidance of stress and anxiety to subjects.

Potential Risks of Social Science Research

Deception/Incomplete Disclosure

The use of deception, crucial to some social science experiments, may cause psychological harm. Incomplete disclosure concerning the nature of the research, if it poses only minimal risks and is followed by adequate debriefing, is usually acceptable by scholarly standards. In reviewing projects where deception of subjects is proposed, IRBs should assure themselves that deception is in fact necessary to elicit the data required for the study and that deception is a common methodology in the kind of research being proposed. Deception does not automatically preclude subjects consent to participation. Wherever possible the IRB should require that subjects receive adequate information regarding the nature of the risks associated with their involvement even though the full details of the procedures may not be explained beforehand.

Privacy/Confidentiality

The IRB should assure itself that every effort is made by the researcher to protect subjects from potential economic loss, legal jeopardy, public embarrassment, etc. IRB review should ensure that steps are taken to protect the privacy of subjects, to ensure the confidentiality of their responses, and to make certain that informed consent has been obtained properly.

Exploitation

The IRB with its ability to review the broad spectrum of research taking place at its institution, should be aware that certain groups and communities can become the favorite subjects for a variety of research projects. The sum total of those projects could expose these communities to negative publicity for behaviors that are likely to be found in other segments of society as well.

Ethical Principles of Research: Problems in Social Science Research

Beyond the risks of psychological harm, economic loss, or personal humiliation the National Commission, in *The Belmont Report,* expressed a desire to see three ethical principles upheld by all researchers.[4] The first is respect for personhood. Individuals should be treated as autonomous agents. Persons with diminished autonomy, particularly those incapable of self-determination, should be protected. The second principle, beneficence, aims at securing individual well-being; that is, maximizing possible benefits as well as minimizing possible harm. The third principle is that of justice in the sense of fairness. Research subjects should not be singled out simply because of their easy availability or their compromised position.[5]

While these principles seem simple enough, there are those who argue that social science research differs sufficiently in content, method, and philosophy from biomedical research that the application of these principles must be viewed differently. "In social psychology, there is often the question of the use of deception in documentary research, the question of privacy of records in interview research, the major problem in protecting confidences. They require different solutions" (Mosteller, 1980). Hospitals are legally responsible to their patients, thus some form of peer review of patient studies makes sense to researchers, but Mosteller argues that the same procedures are inappropriate to social science research. De Sola Pool agrees and adds that a social science researcher's first obligation, not unlike a reporter's, is to the public that he or she informs; i.e., "It is not (the social scientist's) obligation to do no harm to grafters, criminals, bigots and others whom they study. When they give their word for secrecy, of course, they keep it, but otherwise, their job is to expose, not to protect the human."[6] De Sola Pool's argument does not negate the danger associated with research involving less competent individuals who might indeed suffer if investigators pursue their quest for knowledge without prior review.

Seiler and Murtha (1980) outlined some possible unintended consequences of federally-mandated IRB review of social science research:

1. Because social science research does not have the political neutrality we tend to associate with biomedical research, pressure aimed at curtailing politically controversial social science research projects could dominate IRB review.

2. In an effort to curry favor with HHS (and not necessarily out of a concern for human subjects), institutions may be overly conservative in their evaluation of the risks associated with research projects. Certain research activities may be restricted by review boards whose membership is heavily weighted toward administrators, since some administrators are more concerned about protecting their institutions from criticism than about the potential gains associated with high risk research projects.
3. Some institutions may use the review process as a means of controlling "dissidents."
4. Efforts to codify ethical conduct may encourage research subjects to test their rights or to seek compensation through the courts, even in cases where no demonstrable harm has occurred. If this happens, further government regulation of social science research is sure to follow.
5. Increasing bureaucratization of research inevitably increases delays, increases the work load involved in doing research, increases the politicization of scholarly pursuits, and diminishes the control of the researcher over his or her work situation.

Seiler and Murtha (1980) conclude that insofar as federal regulations apply to social research, "they were designed to protect a category of people who never asked for help ... from problems which have never been demonstrated to exist."

The costs and benefits of social research involving human subjects are difficult to calculate. Certainly, subjective judgments dominate such benefit-detriment calculations. The question, then, is also who should make these judgments? The social scientists who argue against case-by-case reviews of their research claim that they are the only ones who can make these judgments. The federal government (and the public) are relying on IRBs to make these difficult calculations.

The Evidence to Date

There have been several studies of IRBs. The most prominent is a 61-institution study conducted for the National Commission by the Survey Research Center at the University of Michigan.[7] Interviews were conducted with more than 2000 researchers and 800 review board members. The interviews included questions concerning the impact of the IRBs over a one-year period ending in 1975.

The Michigan study showed that although IRBs commonly require some changes in proposed research plans (55% of the projects of those surveyed), the rejection of proposals (3%) is rare and 85% of the institutions surveyed had rejected not one project in that year (Gray and Cooke, 1976). The most frequent IRB request is for supplemental information regarding research strategy. The second most frequent request is for modification of consent procedures (usually involving the structure of the actual consent form). (See Table I.)

When asked to assess the impact of the review process on their own research, 4.5% of the respondents indicated that it was a great impediment, 27.5% said they were impeded somewhat, 31% said they were impeded very little, and 37% said they were not impeded at all. In addition, 91% of the respondents viewed the judgments and recommendations of the review boards as sound (Table II). Just over half the respondents stated that the benefits of the IRB process outweighed any difficulties that had arisen.

The Michigan study also indicates that researchers who had once served on an IRB (one-quarter of all those interviewed) were more likely than other researchers to regard as impermissible (i) reuse of data without informed consent, (ii) research of no direct benefit to the subjects, (iii) involvement of children, and (iv) withholding of information from subjects (Gray and Cooke, 1980). Gray states, however, that it is not possible to tell from the Michigan data whether these differences are a result of IRB membership or the result of sample selection.

In February and March of 1980, IRB heads from 10 institutions were interviewed at some length by the authors of this paper.[8] Our goal was to determine the extent to which IRB review of social science research proposals had caused problems for either the researchers or the IRBs.[9] The majority of individuals interviewed believed that social science research could be as harmful to human subjects as biomedical research. While the harm might not involve an assault on the body, they suggested that psychological damage could

Table I. Action Formally Required of the Investigator by the Review Board[ab]

Type of request	Universities (n = 514)	Medical schools (n = 1425)	Hospitals (n = 254)	Mental institutions (n = 101)	Other (n = 95)	All (n = 2389)
More information	33	30	39	28	21	32
Modification:						
Consent form or procedure	19	25	31	14	13	24
Scientific design	—[c]	2	6	8	1	3
Subject selection	—[c]	3	5	7	1	3
Risk or discomfort	3	4	4	7	9	4
Confidentiality	6	2	3	6	6	3
Other	5	3	7	7	9	5
Informal suggestion for modification	13	15	13	19	15	15

[a] From Gray *et al.* (1978), reprinted with permission of the American Academy of Sciences.
[b] Data expressed as percentages.
[c] Less than 9.5% but greater than 0.

Table II. Attitudes of Different Types of Investigators and Review Committee
Members toward Review Procedure and Committees[a,b]

	Review board members			Research investigators		
	Biomedical sciences (n = 370)[c]	Behavioral social sciences (n = 135)[c]	Other (n = 220)[c]	Biomedical sciences (n = 940)[c]	Behavioral and social sciences (n = 395)	Other (n = 180)
The human subjects review procedure has protected the rights and welfare of human subjects—at least to some extent.	99	99	99	99	96	98
The review procedure has improved the quality of scientific research done at this institution—at least to some extent.	78	62	70	69	55	83
The review committee gets into areas that are not appropriate to its function—at least to some extent	39	24	27	50	49	39
The review committee makes judgments that it is not qualified to make—at least to some extent.	28	21	20	43	49	25
The review procedure has impeded the progress of research at this institution—at least to some extent.	26	30	22	43	54	36

[a] From Institutional review boards: Report and recommendations of the National Commission for the Protection of Human Subjects of Biomedical and Behavioral Research, *Federal Register,* Vol. 43, No. 231, November 30, 1978, p. 56191.
[b] Data expressed as percentage agreeing with each statement.
[c] n is approximate since nonresponse varied deed from item to item.

be just as real. The most frequently mentioned concerns of the IRB heads interviewed were invasion of privacy, confidentiality, psychological damage, and deception. The IRB heads indicated that it is the researchers' lack of familiarity with IRB regulations and their propensity to submit too little information concerning their proposed research that causes much of the delay that researchers complain about.

The majority of those interviewed felt that an informed consent document need not be obtained for all research involving human subjects. Most noted that a willingness to be surveyed by telephone or to complete a questionnaire can be presumed to imply consent. They also indicated that research involving unobtrusive observations of public behavior need not require informed consent of the subject. In cases involving crucial deadlines and minimal risks, some IRB heads indicate that they give approval prior to formal IRB review. None of the institutions contacted had a formal IRB appeals process. Investigators may come before the board to discuss their proposals, present additional information, or subsequently modify and resubmit proposals.

Our interviewees revealed that many of the concerns that had prompted the federal government to include nonbiomedical research under the purview of IRBs were appropriate. Invasion of privacy has arisen as an issue in surveys of sexual behavior, studies of grief, studies of illegal immigration, and research involving the use of persons from lists that are not publicly accessible. IRBs, according to our interviewees, have insisted that the subjects of such research be given whatever assistance they need to understand the nature of the research project and their right not to answer questions. The IRB heads interviewed pointed out that problems concerning the confidentiality of data have often been resolved by requiring the further development of coding systems that increase the guarantee of anonymity for participants. Concerns about the economic risks to subjects have arisen in cases involving the study of labor–management relations and illegal immigration. The IRB heads interviewed felt that there were potentially serious risks to subjects involved in such research. Embarrassment and potential psychological damage surfaced as issues in proposed studies of grief and the terminally ill, sex-related studies, studies involving the attitudes of minors, and experiments in which subjects are exposed to unforseen circumstances. Problems of deception were noted as issues in cases of bystander research and studies involving children.

Our interviews suggest that IRB procedures vary with the scale of research activities at each institution. In some universities, academic departments are delegated the responsibility for initial screening in an effort to weed out proposals that present minimal risk or do not involve deception. In the majority of instances, however, all IRB members review each proposal, or at least a condensed version of the proposal.

The results of the Michigan study and our own interviews do not support the contention that IRBs have encroached significantly on the rights of investigators. At the ten universities we studied, we could find only four proposed projects that had been stopped entirely. Two involved highly deceptive situations in which the proposed subjects were children; in both cases the age of the children prevented debriefing. A third study, involving the use of alcohol and drugs, was thought to be risky and of little merit. The fourth involved the use of hypnotism; the capacity of the investigator to carry out the procedure was questioned. In each of these cases, disapproval was clearly based on the need to protect the subjects from unnecessary harm or to protect those who were unable to protect themselves. The Michigan study also found that the only social research projects that were prohibited involved children or cases of serious deception (Gray and Cooke, 1980).

A substantial number of research projects have been modified in response to the recommendations of IRBs. Modifications have taken several forms. Our study indicates that some modifications were requested in an effort to provide more explicit mechanisms for the protection of the anonymity of subjects (e.g., in the case of interviews with illegal immigrants). In studies involving work performance, IRBs sought to ensure the confidentiality of data to protect employees from possible employer retaliation. Other modifications involved the call for more explicit plans for debriefing participants involved in deceptive research and for more complete consent forms. Minor modifications involved the rewording of survey questions and the alteration of survey protocols. A great many researchers were asked to submit additional information more fully explaining the objectives of their studies. Delays have caused some aggravation for investigators, but on the whole, as reported above, the motives of the IRBs have appeared reasonable.

IRB review has, unfortunately, not been treated as a learning process. Few, if any, IRBs have codified uniform standards based on their experiences thus far. Our study indicates that almost all proposals are treated as if nothing like them had ever been reviewed before. In other words, precedent carries little weight. There seems to be a predisposition on the part of many IRBs to search for potential harm, even in proposals that duplicate previously approved projects. The IRB heads interviewed gave no indication of having reached any greater understanding of the special characteristics of social science research. Most institutions have made little, if any, progress in communicating to their members what their IRBs have learned thus far about the risks involved in human subject research or about ways of coping with them. In the absence of such communication, of course, researchers have not been able to do a better job of regulating themselves more effectively.

Additional Concerns of Social Scientists

Definitions. HHS regulations present some continuing definitional problems for social scientists. Research is currently defined as "a systematic investigation designed to develop or contribute to generalizable knowledge.[10] This definition grows out of the proposition that if any activity is submitted for research funding, then it must constitute research. As research methods have expanded, definitions have not been clarified. They remain terribly broad and, as Gray (1980) points out, still leave unanswered the question of whether teaching activities and studies involving description rather than generalization ought to be included (Gray, 1980).

Sensitive Questions. Although previous regulations made no mention of whether IRBs should be concerned about sensitive questions being asked in the course of research, some IRBs have considered the sensitivity of questions as a factor in assessing the risks associated with social science research proposals. Collection of sensitive information is a factor in determining whether research will be covered by, or exempted from, IRB review. Nonsocial scientists sometimes assume that respondents have no defense against sensitive questions; many survey researchers feel that adult subjects can decide for themselves whether or not to answer questions of a particularly sensitive nature (Gray, 1980).

Exemptions. Categorical exemptions or expedited review for certain types of low risk research have been provided for by HHS regulations along with a waiver of the informed consent requirement, where applicable. This should permit IRBs to spend more time reviewing proposals that pose substantial harm to human subjects. Whether IRBs agree to such permit exemptions may be a matter of institutional policy. Some IRBs may decide to take an overly conservative approach.

Consent Forms. The flexibility of consent form procedures should ease a number of problems. Although categorical exemptions remove some difficulties previously noted by social scientists, new forms of low risk research not presently covered by the final regulations are sure to develop. These would then be subject to stringent informed consent procedures. The solution, of course, is for the exemption list to be amended periodically. Robertson (1979) noted also the need to improve the readability and completeness of forms for obtaining consent. He suggests that many subjects are not given sufficient time to consider whether or not they want to participate.

Appeals Procedures. HHS regulations do not require that an IRB establish an appeals process for negative IRB decisions. Although few projects are being rejected outright, researchers have no recourse at present except legal action. Some institutions have established departmental IRBs as well as institution-wide boards. Investigators have the right to meet with IRB members in an effort to justify their proposed research, but once turned down by the board itself they have no other avenue of appeal.

Other Ways of Thinking about the Issue

The adoption of regulations concerning research on human subjects is probably an outcome rather than a cause of growing concern about bioethical and social issues in the United States. The considerations emphasized and decisions made by IRBs today differ markedly from those we would probably have seen two decades ago. Public concern about the behavior of professionals and scientists has increased along with the realization that judgment rather than expertise dominates many of the decisions that professionals make on our behalf. Federal regulation and calls for peer review of professional behavior are a response to public concern. Social scientists, uneasy about the involvement of government in the regulation of academic research, would do well to consider the emergence of IRBs in the broader context of the consumer movement rather than under the narrower heading of increased government interference in academic life.

The IRB process has helped to improve research practices. It has made researchers more aware of ethical issues and (because of fear about losing government financial support) ensured that federally funded research meets prevailing professional norms of acceptability for the utilization of human subjects. The increased sensitivity of institutions to the need for peer review and concern about the risks to the community at large have caused some overreactions. Social science research projects have, on occasion, been subject to overly stringent review procedures. There are some documented instances in which the demands on investigators have been totally unreasonable—mostly as a result of the IRBs' inability to distinguish between the needs of biomedical and nonbiomedical research and the risks that each involves. The social science community may also have overreacted, at least in part, to the atypical cases.

The red flag of federal regulation has been waved. Claims of prior restraint on free speech have been made. The peer review process appears to be working reasonably well, although improvements could be made.

Reference Notes

1. The Public Health Service, PPO number 129, February 8, 1966; revised July 1, 1966. The major medical experiment was conducted in 1964 at the Jewish Chronic Disease Hospital where researchers injected live cancer cells beneath the skin of nonconsenting geriatric patients.
2. *Belmont Report, Ethical Principles and Guidelines for the Protection of Human Subjects of Research,* Vol. II, Appendix, p. 18-1, DHEW Publication No. (OS) 78-0014.
3. The most outspoken critic of the proposed regulations is Dr. Ithiel de Sola Pool, Sloan Professor of Political Science at the Massachusetts Institute of Technology.
4. The National Commission for the Protection of Human Subjects of Biomedical and Behavioral Research, *The Belmont Report, Ethical Principles and Guidelines for the Protection of Human Subjects of Research,* DHEW Publication No. (OS) 78-0012, Washington, D.C., 1978. (Published in the *Federal Register,* April 18, 1979, Vol. 44, No. 76, pp. 23192–23197.)
5. *Ibid.,* p. 23194.
6. Ithiel de Sola, from a copy of testimony before the President's Commission for the Study of Ethical Problems in Medicine and Biomedical and Behavioral Research, July 12, 1980, p. 3.
7. Survey Research Center, University of Michigan, 1978 *Research involving human subjects:* in *Reports and Recommendations of the National Commission for the Protection of Human Subjects of Biomedical and Behavioral Research,* DHEW Publication No. (OS) 78-009.
8. The Institutions asked to participate in our study were University of Illinois, The Johns Hopkins University, University of Texas—Austin, University of Southern California—Los Angeles, University of Pennsylvania, University of Iowa, University of Washington, Colum-

bia University, Harvard University, Dartmouth College, University of North Carolina, Georgia Tech., MIT, University of Michigan, and Duke University. Ten of the chairpersons contacted agreed to be interviewed by telephone. The average length of each interview was one hour.

9. Our survey instrument was reviewed and modified by the IRB at MIT. We asked a series of open-ended questions aimed at discovery; (i) the extent to which IRBs review social science research, (ii) whether review procedures are the same for biomedical and social research, (iii) whether social research is viewed as potentially as harmful as other types of research, and (iv) the extent to which specific issues such as invasion of privacy, job or economic detriment, embarrassment or psychological damage, liable or slander, deception, censorship or infringement of academic freedom, and prior restraint had been raised.

10. *Federal Register,* January 26, 1981, Vol. 46, 102(e).

References

Caplan, A. L., 1979, HEW's painless way to review human research, *The New York Times,* 1979(December 27):A22.

Gray, B. H., 1979, Human subjects review committees and social research, in: *Federal Regulations, Ethical Issues and Social Research* (M. L. Wax and J. Cassell, eds.) pp. 43–58, Westview Press, Boulder, Colorado.

Gray, B. H., 1980, Social research and the proposed DHEW regulations, *IRB* 2(1):1–5, 12.

Gray, B. H., and Cooke, R. A., 1980, The impact of institutional review boards on research, *Hastings Cent. Rep.* 10(1):36–41.

Gray, B. H., Cooke, R. A., and Tannenbaum, A. S., 1978, Research involving human subjects, *Science* 201:1094–1101.

Mosteller, F., 1980, Regulation of Social Research, *Science* 208:1.

Pool, I. S., 1979, Prior restraint, *The New York Times* 1979(December 16):64.

Robertson, J. A., 1979, Ten ways to improve IRBs, *Hastings Cent. Rep.* 9(1):29–33.

Seiler, L. H. and Murtha, J. M., 1979, Federal regulation of social research, in: *Freedom at Issue* (reprint) Freedom House, New York.

Seiler, L. H. and Murtha, J. M., 1980, Federal regulation of social research using human subjects: A critical assessment, *Am. Sociologist* 15(3):146–155.

Smith, M. B., 1979, Some Perspectives on ethical/political issues in social science research, in: *Federal Regulations, Ethical Issues and Social Research* (M. L. Wax and Joan Cassell, eds.), pp. 11–21, Westview Press, Boulder Colorado.

Bibliography

Barber, B., 1980, Regulation and the professionals, *Hastings Cent. Rep.* 10(1):34–36.

Frankel, M. S., 1975, The development of policy guidelines governing human experimentation in the United States, *Ethics Sci, Med.* 2:43–59.

Gray, B. H., Cooke, R. A. and Tannenbaum, A. S., 1978, Research involving human subjects, *Science* 201:1094–1101.

Pool, I. S., 1980, Testimony before the President's Commission for the study of ethical problems in medicine and biomedical and behavioral research, July, 12, 1980.

Pool, I. S., 1980, Letter to Morris Abrams and Alexander Capron, members of the President's Commission, concerning documentation of testimony, July 29, 1980.

Robertson, J. A., The law of institutional review boards, *UCLA Law Rev.* 26:484–549.

Tropp, R. A., What problems are raised when the current DHEW regulation on protection of human subjects is applied to social science research, Belmont Report, Appendix II, pp. 18-1 to 18-17.

U.S. Code Congressional and Administrative News, 1974, The National Research Act, PL 93-348; pp. 3635-3689.

Federal Register, Title 45, Part 46—Protection of Human Subjects, Vol. 39 (No.) 105, May 30, 1974, pp. 18914-18920.

Federal Register, Title 45, Part 46—Protection of Human Subjects, Vol. 46, No. 16, January 26, 1981, pp. 8366-8392.

Veatch, R. M., 1979, The national commission of IRBs: An evolutionary approach, *Hastings Cent. Rep.* 9(1):22-28.

Wax, M. L., and Cassell, J. (eds.), 1979, *Federal Regulations, Ethical Issues and Social Research,* Westview Press, Boulder, Colorado.

ANNOTATED
BIBLIOGRAPHY

Annotated Bibliography

Ackerman, T. F., 1979, Fooling ourselves with child autonomy and assent in nontherapeutic clinical research, *Clin. Res.* **27**:345.

A dissenting opinion on the National Commission recommendation for an "assent" procedure.

Annas, G., 1978, Informed consent, *Annu. Rev. Med.* **29**:9.

A widely quoted article dealing with clinical consent rather than research: includes comments on various state laws.

Barber, B., Lally, J. J., Makarushka, J., and Sullivan, D., 1973, Research on Human Subjects, Russell Sage Foundation, New York.

A monograph on the problems of social control in medical experimentation.

Barber, B., 1980, Informed Consent in Medical Therapy and Research, Rutgers University Press, New Brunswick, New Jersey.

An advanced treatise in which the author, a noted sociologist, analyzes informed consent in terms of Western society's value system, the authority relationship between physician and subject, problems in communication, etc.

Bonchek, L., 1979, Are randomized trials appropriate for evaluating new operations, *N. Engl. J. Med.* **301**:44.

Comment on a difficult problem in surgical research.

Brown, J. H. U., 1978, Management of an institutional review board for the protection of human subjects, *SRA Journal* (Society of Research Administrators) **Summer**:5.

Comments on IRB function from the point of view of a research administrator.

Brown, J. H. U., Schoenfeld, L., and Allan, P. W., 1979, The costs of an institutional review board, *J. Med. Educ.* **54**:294.

At the University of Texas Health Science Center in San Antonio, it cost $100,000 to review 850 protocols, the costs all being borne by the institution.

Cowan, D. H., 1975, Human experimentation: the review process in practice, *Case West. Reserve Law Rev.* **25**:533.

A description of the system in place at a major university in 1975, with discussion of risk/benefit ratio, informed consent, and problems of IRB review.

Curran, W. J., 1979, Compensation for injured research subjects. *N. Engl. J. Med.* **301**:648.

Comment on the "interim final" regulation of 1979.

Gray, B. H., 1977, The functions of human subjects review committees, *Am. J. Psychiat.* **134**:9.
A discussion of several "latent" functions of IRBs, vis à vis protection of institutions rather than subjects, judgment of research by social impact, and determination of "community acceptability."

Gray, B. H., Cooke, R. A., and Tannenbaum, A. S., 1978, Research involving human subjects, *Science* **201**:1094.
A study of the performance of 61 IRBs performed during 1974–1975, revealing that many IRBs at that time were quite deficient in their responsibilities ·with particular respect to poor review of unacceptable consent forms and lack of impact on the consent process.

Hershey, N., and Miller, R., 1976, Human Experimentation and the Law, Aspen Systems Corporation, Germantown, Md.
A monograph of IRB functioning written before the report of the National Commission and the subsequent changes in regulations; provides an overview of the historical background, of IRB concerns, and of applicable law.

Huff, T. A., 1979, The IRB as deputy sheriff: Proposed FDA regulation of the institutional review board, *Clin. Res.* **27**:103.
Comment on the disparity between the IRB as a "review" group and the attempt by the FDA to make it a "regulator," an important distinction.

Lally, J. J., 1978, The making of the compassionate physician/investigator, *Ann. Am. Acad. Pol. Soc. Sci.* **437**:86.
An overview of the training of research physicians with respect to development of ethical standards of concern for human subjects.

Lebacqz, K., and Levine, R. J., 1977, Respect for persons and informed consent to participate in research, *Clin. Res.* **25**:101.
A sophisticated ethical analysis of informed consent based on the principle of respect for persons, in which it is argued that increased IRB review and monitoring of research, while protecting people from harm, may intrude on their right to self-determination.

Levine, C. (ed.), 1978 to date, *IRB: A review of Human Subjects Research*, Hastings Center, Hastings-on-Hudson, New York.
A periodical (ten issues/year) dealing with many aspects of IRB procedure, ethical concerns, controversies, etc.; required reading for all IRB members and administrators.

Levine, R. J., 1976, The institutional review board, in: *Report and Recommendations of the National Commission for the Protection of Human Subjects of Biomedical and Behavioral Research* (appendix).
A description of the IRB at Yale University, with details of its procedures, plus discussions of important issues such as opening IRB meetings to the public, educational activities of the IRB, monitoring, and communications problems.

Makarushka, J. L., and McDonald, R. D., 1979, Informed consent, research, and geriatric patients: The responsibility of institutional review committees, *Gerontologist* **19**:61.

Comment on special considerations applicable to research on geriatric patients.

May, W., 1975, The composition and function of ethical committees, *J. Med. Ethics* **1**:23.

A comparison of IRB function in the U.K. and the U.S., with comments on scientific review.

Melmon, K., Grossman, M. and Curtis, M. R., 1970, Emerging assets and liabilities of a committee on human welfare and experimentation, *N. Engl. J. Med.* **282**:427.

Of historical interest, an account of the activities of an early IRB at the University of California San Francisco, highlighting problems of acceptance by the scientific staff of IRB activity.

National Commission for the Protection of Human Subjects of Biomedical and Behavioral Research, 1978, Report and Recommendations on Institutional Review Boards, Superintendent of Documents, Washington, D.C.

The pivotal document in IRB evolution which led to many of the changes proposed in the *Federal Register* of August 14, 1979; includes a series of recommendations (with discussion), plus chapters on ethical guidelines, performance of IRBs, legal aspects, and federal policies. Accompanied by an appendix of supporting documents, including an important paper by Levine *(vide supra)*.

Park, L. C., Covi, L., and Uhlenhuth, E. H., 1967, Effects of informed consent on research patients and study results, *J. Nerv. Mental Dis.* **145**:349.

Although the consent process used in 1967 was different from that of today, these investigators found that informed consent was an adjunct to a drug study on psychiatric patients rather than a hindrance.

Richmond, D. E., 1977, Auckland hospital ethical committee: The first three years, *N. Z. Med. J.* **86**:10.

An account of an IRB abroad, highlighting the similarity of problems worldwide.

Roginsky, M. S., and Handley, A., 1978, Ethical implications of withdrawal of experimental drugs at the conclusion of phase III trials, *Clin. Res.* **26**:384.

A discussion of the ethical dilemma faced by both patient and physician when successful drug therapy must be stopped at the end of a trial.

Sorenson, A. A., 1978, A sociological study of informed consent in a university hospital: Problems with the institutional review board, *Clin. Res.* **26**:1.

A disturbing account of an encounter between a sociologist and an IRB in which the former wanted to study the consent process *per se* and was thwarted on tenuous grounds by the committee.

Woodward, W. E., 1979, Informed consent of volunteers: A direct measurement of comprehension and retention of information, *Clin. Res.* **27**:248.

An interesting study in which volunteers for a cholera vaccine trial were given detailed information prior to participation and then tested for comprehension and recall.

APPENDICES

HISTORICAL DOCUMENTS

The Nuremberg Code

Permissible Medical Experiments

The great weight of the evidence before us is to the effect that certain types of medical experiments on human beings, when kept within reasonably well-defined bounds, conform to the ethics of the medical profession generally. The protagonists of the practice of human experimentation justify their views on the basis that such experiments yield results for the good of society that are unprocurable by other methods or means of study. All agree, however, that certain basic principles must be observed in order to satisfy moral, ethical and legal concepts:

1. The voluntary consent of the human subject is absolutely essential. This means that the person involved should have legal capacity to give consent; should be so situated as to be able to exercise free power of choice, without the intervention of any element of force, fraud, deceit, duress, over-reaching, or other ulterior form of constraint or coercion; and should have sufficient knowledge and comprehension of the elements of the subject matter involved as to enable him to make an understanding and enlightened decision. This latter element requires that before the acceptance of an affirmative decision by the experimental subject there should be made known to him the nature, duration and purpose of the experiment; the method and means by which it is to be conducted; all inconveniences and hazards reasonably to be expected; and the effects upon his health or person which may possibly come from his participation in the experiment.

The duty and responsibility for ascertaining the quality of the consent rests upon each individual who initiates, directs, or engages in the experiment. It is a personal duty and responsibility which may not be delegated to another with impunity.

2. The experiment should be such as to yield fruitful results for the good of society, unprocurable by other methods or means of study, and not random and unnecessary in nature.

3. The experiment should be so designed and based on the results of animal experimentation and a knowledge of the natural history of the disease or other problem under study that the anticipated results will justify the performance of the experiment.

4. The experiment should be so conducted as to avoid all unnecessary physical and mental suffering and injury.

5. No experiment should be conducted where there is a prior reason to believe that death or disabling injury will occur; except, perhaps, in those experiments where the experimental physicians also serve as subjects.

6. The degree of risk to be taken should never exceed that determined by the humanitarian importance of the problem to be solved by the experiment.

7. Proper preparations should be made and adequate facilities provided to protect the experimental subject against even remote possibilities of injury, disability, or death.

8. The experiment should be conducted only by scientifically qualified persons. The highest degree of skill and care should be required through all stages of the experiment of those who conduct or engage in the experiment.

9. During the course of the experiment the human subject should be at liberty to bring the experiment to an end if he has reached the physical or mental state where continuation of the experiment seems to him to be impossible.

10. During the course of the experiment the scientist in charge must be prepared to terminate the experiment at any stage, if he has probable cause to believe, in the exercise of the good faith, superior skill and careful judgment required of him that a continuation of the experiment is likely to result in injury, disability, or death to the experimental subject.

Of the ten principles which have been enumerated our judicial concern, of course, is with those requirements which are purely legal in nature--or which at least are so closely related to matters legal that they assist us in determining criminal culpability and punishment. To go beyond that point would lead us into a field that would be beyond our sphere of competence. However, the point need not be labored. We find from the evidence that in the medical experiments which have been proved, these ten principles were much more frequently honored in their breach than in their observance. Many of the concentration camp inmates who were the victims of these atrocities were citizens of countries other than the German Reich. They were non-German nationals, including Jews and "asocial persons," both prisoners of war and civilians, who had been imprisoned and forced to submit to these tortures and barbarities without so much as a semblance of trial. In every single instance appearing in the record, subjects were used who did not consent to the experiments; indeed, as to some of the experiments, it is not even contended by the defendants that the subjects occupied the status of volunteers. In no case was the experimental subject at liberty of his own free choice to withdraw from any experiment. In many cases experiments were performed by unqualified persons; were conducted at random for no adequate scientific reason, and under revolting physical conditions. All of the experiments were conducted with unnecessary suffering and injury and but very little, if any, precautions, were taken to protect or safeguard the human subjects from the possibilities of injury, disability, or death. In every one of the experiments the subjects experienced extreme pain or torture, and in most of them they suffered permanent injury, mutilation, or death, either as a direct result of the experiments or because of lack of adequate follow-up care.

Obviously all of these experiments involving brutalities, tortures, disabling injury, and death were performed in complete disregard of international conventions, the laws and customs of war, the general principles of criminal law as derived from the criminal laws of all civilized nations, and Control Council Law No. 10. Manifestly human experiments under such conditions are contrary to "the principles of the law of nations as they result from the usages established among civilized peoples, from the laws of humanity, and from the dictates of public conscience."

Whether any of the defendants in the dock are guilty of these atrocities is, of course, another question . . . .

Declaration of Helsinki

Recommendations Guiding Medical Doctors
in
Biomedical Research Involving Human Subjects

Adopted by the 18th World Medical Assembly, Helsinki, Finland, 1964, and as
revised by the 29th World Medical Assembly, Tokyo, Japan, 1975

INTRODUCTION

It is the mission of the medical doctor to safeguard the health of the people.
His or her knowledge and conscience are dedicated to the fulfillment of this
mission.

The Declaration of Geneva of the World Medical Association binds the doctor
with the words, "The health of my patient will be my first consideration,"
and the International Code of Medical Ethics declares that, "Any act or advice
which could weaken physical or mental resistance of a human being may be used
only in his interest."

The purpose of biomedical research involving human subjects may be to improve
diagnostic, therapeutic, and prophylactic procedures and the understanding of
the aetiology and pathogenesis of disease.

In current medical practice most diagnostic, therapeutic or prophylactic procedures involve hazards. This applies a _fortiori_ to biomedical research.

Medical progress is based on research which ultimately must rest in part on
experimentation involving human subjects.

In the field of biomedical research a fundamental distinction must be recognized between medical research in which the aim is essentially diagnostic or
therapeutic for a patient, and medical research, the essential object of
which is purely scientific and without direct diagnostic or therapeutic value
to the person subject to the research.

Special caution must be exercised in the conduct of research which may affect
the environment, and the welfare of animals used for research must be respected.

Because it is essential that the results of laboratory experiments be applied
to human beings to further scientific knowledge and to help suffering humanity,
the World Medical Association has prepared the following recommendations as
a guide to every doctor in biomedical research involving human subjects.
They should be kept under review in the future. It must be stressed that the
standards as drafted are only a guide to physicians all over the world.
Doctors are not relieved from criminal, civil, and ethical responsibilities
under the laws of their own countries.

I. Basic Principles

1. Biomedical research involving human subjects must conform to generally
accepted scientific principles and should be based on adequately performed
laboratory and animal experimentation and on a thorough knowledge of the
scientific literature.

2. The design and performance of each experimental procedure involving human
subjects should be clearly formulated in an experimental protocol which should
be transmitted to a specially appointed independent committee for consideration, comment and guidance.

3. Biomedical research involving human subjects should be conducted only by scientifically qualified persons and under the supervision of a clinically competent medical person. The responsibility for the human subject must always rest with a medically qualified person and never rest on the subject of the research, even though the subject has given his or her consent.

4. Biomedical research involving human subjects cannot legitimately be carried out unless the importance of the objective is in proportion to the inherent risk to the subject.

5. Every biomedical research project involving human subjects should be preceded by careful assessment of predictable risks in comparison with foreseeable benefits to the subject or to others. Concern for the interests of the subject must always prevail over the interests of science and society.

6. The right of the research subject to safeguard his or her integrity must always be respected. Every precaution should be taken to respect the privacy of the subject and to minimize the impact of the study on the subject's physical and mental integrity and on the personality of the subject.

7. Doctors should abstain from engaging in research projects involving human subjects unless they are satisfied that the hazards involved are believed to be predictable. Doctors should cease any investigation if the hazards are found to outweigh the potential benefits.

8. In publication of the results of his or her research, the doctor is obliged to preserve the accuracy of the results. Reports of experimentation not in accordance with the principles laid down in this Declaration should not be accepted for publication.

9. In any research on human beings, each potential subject must be adequately informed of the aims, methods, anticipated benefits and potential hazards of the study and the discomfort it may entail. He or she should be informed that he or she is at liberty to abstain from participation in the study and that he or she is free to withdraw his or her consent to participation at any time. The doctor should then obtain the subject's freely-given informed consent, preferably in writing.

10. When obtaining informed consent for the research project the doctor should be particularly cautious if the subject is in a dependent relationship to him or her or may consent under duress. In that case the informed consent should be obtained by a doctor who is not engaged in the investigation and who is completely independent of this official relationship.

11. In case of legal incompetence, informed consent should be obtained from the legal guardian in accordance with national legislation. Where physical or mental incapacity makes it impossible to obtain informed consent, or when the subject is a minor, permission from the responsible relative replaces that of the subject in accordance with national legislation.

12. The research protocol should always contain a statement of the ethical considerations involved and should indicate that the principles enunciated in the present Declaration are complied with.

II. Medical Research Combined with Professional Care (Clinical Research)

1. In the treatment of the sick person, the doctor must be free to use a new diagnostic and therapeutic measure, if in his or her judgment it offers hope of saving life, reestablishing health or alleviating suffering.

2. The potential benefits, hazards and discomfort of a new method should be weighed against the advantages of the best current diagnostic and therapeutic methods.

3. In any medical study, every patient - including those of a control group, if any -- should be assured of the best proven diagnostic and therapeutic methods.

4. The refusal of the patient to participate in a study must never interfere with the doctor-patient relationship.

5. If the doctor considers it essential not to obtain informed consent, the specific reasons for this proposal should be stated in the experimental protocol for transmission to the independent committee (I,2).

6. The doctor can combine medical research with professional care, the objective being the acquisition of new medical knowledge, only to the extent that medical research is justified by its potential diagnostic or therapeutic value for the patient.

III. Non-Therapeutic Biomedical Research Involving Human Subjects
     (Non-clinical biomedical research)

1. In the purely scientific application of medical research carried out on a human being, it is the duty of the doctor to remain the protector of the life and health of that person on whom biomedical research is being carried out.

2. The subjects should be volunteers -- either healthy persons or patients for whocm the experimental design is not related to the patient's illness.

3. The investigator or the investigating team should discontinue the research if in his/her or their judgment it may, if continued, be harmful to the individual.

4. In research on man, the interest of science and society should never take precedence over the considerations related to the wellbeing of the subject.

FEDERAL REGULATIONS

HHS Proposed Regulations

Except when otherwise provided by federal, state or local law, information in the records or in the possession of an institution acquired in connection with an activity covered by these regulations which refers to or can be identified with a particular subject, may not be disclosed except: (a) With the consent of the subject or his legally authorized representative; or (b) as may be necessary for the Secretary to carry out his responsibilities. (44 FR 47698)

Public Comment: Fourteen commentators addressed the issues of the privacy of subjects and the confidentiality of information pertaining to them. A majority of those who commented requested deletion or at least modification of this requirement.

HHS Response: The federal government and some states have statutes which provide for the privacy of human subjects and the confidentiality of information pertaining to them. However, few of these laws provide absolute protections. Consequently, it is inappropriate to require institutions to give assurances of privacy and confidentiality which they may not be able to honor in all circumstances.

HHS Decision: The regulations do not have specific requirements describing how personal information must be maintained or to whom it may be disclosed. However, IRBs will be required to determine that, where appropriate, there are adequate provisions to protect the privacy of subjects and to maintain the confidentiality of data (§ 46.111(a)(7)). Confidentiality provisions should meet reasonable standards for protection of privacy and comply with applicable laws. Reasonable protection might in some instances include legal protection available upon application (such as the immunity from legal process of certain drug and alcohol abuse and mental health research subject data under sec. 303 of the PHS Act). In addition, the informed consent provision of the regulations (§ 46.116) requires disclosure to each subject of the extent to which confidentiality of records identifying the subject will be maintained.

The Following Sections of the Regulations Were not Controversial and Were Adopted as Proposed

Section 46.119   Research Undertaken Without the Intention of Involving Human Subjects.
Section 46.120   Evaluation and Disposition of Applications and Proposals.
Section 46.122   Use of Federal Funds.
Section 46.124   Conditions.

Dated December 12, 1980.
Julius B. Richmond,
Assistant Secretary for Health and Surgeon General
Approved January 13, 1981
Patricia Roberts Harris,
Secretary

Accordingly, Part 46 of 45 CFR is amended below by:

§ 46.205   [Amended]

1. Amending § 46.205(b) by changing the reference in the eighth line from "§ 46.115" to "§ 46.120."

§ 46.304   [Amended]

2. Amending § 46.304 by changing the reference in the second line from "§ 46 106" to "§ 46.107."

Subparts A and D [Removed]

3. Removing Subparts A and D and adding the following new Subpart A.

Subpart A—Basic HHS Policy for Protection of Human Research Subjects

Sec.
46.101   To what do these regulations apply?
46.102   Definitions
46.103   Assurances.
46.104   Section reserved.
46.105   Section reserved.
46.106   Section reserved.
46.107   IRB membership.
46.108   IRB functions and operations.
46.109   IRB review of research.
46.110   Expedited review procedures for certain kinds of research involving no more than minimal risk, and for minor changes in approved research.
46.111   Criteria for IRB approval of research.
46.112   Review by institution.
46.113   Suspension or termination of IRB approval of research.
46.114   Cooperative research.
46.115   IRB records.
46.116   General requirements for informed consent.
46.117   Documentation of informed consent.
46.118   Applications and proposals lacking definite plans for involvement of human subjects.
46.119   Research undertaken without the intention of involving human subjects.
46.120   Evaluation and disposition of applications and proposals.
46.121   Investigational new drug or device 30-day delay requirement.
46.122   Use of federal funds.
46.123   Early termination of research funding; evaluation of subsequent applications and proposals.
46.124   Conditions.

Authority: 5 U.S.C. 301; sec. 474(a), 88 Stat. 352 (42 U.S.C. 2891–3(a)).

§ 46.101   To what do these regulations apply?

(a) Except as provided in paragraph (b) of this section, this subpart applies to all research involving human subjects conducted by the Department of Health and Human Services or funded in whole or in part by a Department grant, contract, cooperative agreement or fellowship.

(1) This includes research conducted by Department employees, except each Principal Operating Component head may adopt such nonsubstantive, procedural modifications as may be appropriate from an administrative standpoint.

(2) It also includes research conducted or funded by the Department of Health and Human Services outside the United States, but in appropriate circumstances, the Secretary may, under paragraph (e) of this section, waive the applicability of some or all of the requirements of these regulations for research of this type.

(b) Research activities in which the only involvement of human subjects will be in one or more of the following categories are exempt from these regulations unless the research is covered by other subparts of this part:

(1) Research conducted in established or commonly accepted educational settings, involving normal educational practices, such as (i) research on regular and special education instructional strategies, or (ii) research on the effectiveness of or the comparison among instructional techniques, curricula, or classroom management methods.

(2) Research involving the use of educational tests (cognitive, diagnostic, aptitude, achievement), if information taken from these sources is recorded in such a manner that subjects cannot be identified, directly or through identifiers linked to the subjects.

(3) Research involving survey or interview procedures, except where all of the following conditions exist: (i) Responses are recorded in such a manner that the human subjects can be identified, directly or through identifiers linked to the subjects, (ii) the subject's responses, if they became known outside the research, could reasonably place the subject at risk of criminal or civil liability or be damaging to the subject's financial standing or employability, and (iii) the research deals with sensitive aspects of the subject's own behavior, such as illegal conduct, drug use, sexual behavior, or use of alcohol. All research involving survey or interview procedures is exempt, without exception, when the respondents are elected or appointed public officials or candidates for public office.

(4) Research involving the observation (including observation by participants) of public behavior, except where all of the following conditions exist: (i) Observations are recorded in such a

235

manner that the human subjects can be identified, directly or through identifiers linked to the subjects, (ii) the observations recorded about the individual, if they became known outside the research, could reasonably place the subject at risk of criminal or civil liability or be damaging to the subject's financial standing or employability, and (iii) the research deals with sensitive aspects of the subject's own behavior such as illegal conduct, drug use, sexual behavior, or use of alcohol.

(5) Research involving the collection or study of existing data, documents, records, pathological specimens, or diagnostic specimens, if these sources are publicly available or if the information is recorded by the investigator in such a manner that subjects cannot be identified, directly or through identifiers linked to the subjects.

(c) The Secretary has final authority to determine whether a particular activity is covered by these regulations.

(d) The Secretary may require that specific research activities or classes of research activities conducted or funded by the Department, but not otherwise covered by these regulations, comply with some or all of these regulations.

(e) The Secretary may also waive applicability of these regulations to specific research activities or classes of research activities, otherwise covered by these regulations. Notices of these actions will be published in the Federal Register as they occur.

(f) No individual may receive Department funding for research covered by these regulations unless the individual is affiliated with or sponsored by an institution which assumes responsibility for the research under an assurance satisfying the requirements of this part, or the individual makes other arrangements with the Department.

(g) Compliance with these regulations will in no way render inapplicable pertinent federal, state, or local laws or regulations.

(h) Each subpart of these regulations contains a separate section describing to what the subpart applies. Research which is covered by more than one subpart shall comply with all applicable subparts.

§ 46.102 Definitions.

(a) "Secretary" means the Secretary of Health and Human Services and any other officer or employee of the Department of Health and Human Services to whom authority has been delegated.

(b) "Department" or "HHS" means the Department of Health and Human Services.

(c) "Institution" means any public or private entity or agency (including federal, state, and other agencies).

(d) "Legally authorized representative" means an individual or judicial or other body authorized under applicable law to consent on behalf of a prospective subject to the subject's participation in the procedure(s) involved in the research.

(e) "Research" means a systematic investigation designed to develop or contribute to generalizable knowledge. Activities which meet this definition constitute "research" for purposes of these regulations, whether or not they are supported or funded under a program which is considered research for other purposes. For example, some "demonstration" and "service" programs may include research activities.

(f) "human subject" means a living individual about whom an investigator (whether professional or student) conducting research obtains (1) data through intervention or interaction with the individual, or (2) identifiable private information. "Intervention" includes both physical procedures by which data are gathered (for example, venipuncture) and manipulations of the subject or the subject's environment that are performed for research purposes. "Interaction" includes communication or interpersonal contact between investigator and subject. "Private information" includes information about behavior that occurs in a context in which an individual can reasonably expect that no observation or recording is taking place, and information which has been provided for specific purposes by an individual and which the individual can reasonably expect will not be made public (for example, a medical record). Private information must be individually identifiable (i.e., the identity of the subject is or may readily be ascertained by the   • investigator or associated with the information) in order for obtaining the information to constitute research involving human subjects.

(g) "Minimal risk" means that the risks of harm anticipated in the proposed research are not greater, considering probability and magnitude, than those ordinarily encountered in daily life or during the performance of routine physical or psychological examinations or tests.

(h) "Certification" means the official notification by the institution to the Department in accordance with the requirements of this part that a research

project or activity involving human subjects has been reviewed and approved by the Institutional Review Board (IRB) in accordance with the approved assurance on file at HHS. (Certification is required when the research is funded by the Department and not otherwise exempt in accordance with § 46.101(b)).

§ 46.103 Assurances.

(a) Each institution engaged in research covered by these regulations shall provide written assurance satisfactory to the Secretary that it will comply with the requirements set forth in these regulations.

(b) The Department will conduct or fund research covered by these regulations only if the institution has an assurance approved as provided in this section, and only if the institution has certified to the Secretary that the research has been reviewed and approved by an IRB provided for in the assurance, and will be subject to continuing review by the IRB. This assurance shall at a minimum include:

(1) A statement of principles governing the institution in the discharge of its responsibilities for protecting the rights and welfare of human subjects of research conducted at or sponsored by the institution, regardless of source of funding. This may include an appropriate existing code, declaration, or statement of ethical principles, or a statement formulated by the institution itself. This requirement does not preempt provisions of these regulations applicable to Department-funded research and is not applicable to any research in an exempt category listed in § 46.101.

(2) Designation of one or more IRBs established in accordance with the requirements of this subpart, and for which provisions are made for meeting space and sufficient staff to support the IRB's review and recordkeeping duties.

(3) A list of the IRB members identified by name; earned degrees; representative capacity; indications of experience such as board certifications, licenses, etc., sufficient to describe each member's chief anticipated contributions to IRB deliberations; and any employment or other relationship between each member and the institution; for example: full-time employee, part-time employee, member of governing panel or board, stockholder, paid or unpaid consultant. Changes in IRB membership shall be reported to the Secretary. [1]

[1] Reports should be filed with the Office for Protection from Research Risks, National Institutes of Health, Department of Health and Human Services, Bethesda, Maryland 20205.

8388   **Federal Register** / Vol. 46, No. 16 / Monday, January 26, 1981 / Rules and Regulations

(4) Written procedures which the IRB will follow (i) for conducting its initial and continuing review of research and for reporting its findings and actions to the investigator and the institution; (ii) for determining which projects require review more often than annually and which projects need verification from sources other than the investigators that no material changes have occurred since previous IRB review; (iii) for insuring prompt reporting to the IRB of proposed changes in a research activity, and for insuring that changes in approved research, during the period for which IRB approval has already been given, may not be initiated without IRB review and approval except where necessary to eliminate apparent immediate hazards to the subject; and (iv) for insuring prompt reporting to the IRB and to the Secretary [1] of unanticipated problems involving risks to subjects or others.

(c) The assurance shall be executed by an individual authorized to act for the institution and to assume on behalf of the institution the obligations imposed by these regulations, and shall be filed in such form and manner as the Secretary may prescribe.

(d) The Secretary will evaluate all assurances submitted in accordance with these regulations through such officers and employees of the Department and such experts or consultants engaged for this purpose as the Secretary determines to be appropriate. The Secretary's evaluation will take into consideration the adequacy of the proposed IRB in light of the anticipated scope of the institution's research activities and the types of subject populations likely to be involved, the appropriateness of the proposed initial and continuing review procedures in light of the probable risks, and the size and complexity of the institution.

(e) On the basis of this evaluation, the Secretary may approve or disapprove the assurance, or enter into negotiations to develop an approvable one. The Secretary may limit the period during which any particular approved assurance or class of approved assurances shall remain effective or otherwise condition or restrict approval.

(f) Within 60 days after the date of submission to HHS of an application or proposal, an institution with an approved assurance covering the proposed research shall certify that the application or proposal has been reviewed and approved by the IRB. Other institutions shall certify that the application or proposal has been

---

[1] Reports should be filed with the Office for Protection from Research Risks, National Institutes of Health, Department of Health and Human Services, Bethesda, Maryland 20205.

---

approved by the IRB within 30 days after receipt of a request for such a certification from the Department. If the certification is not submitted within these time limits, the application or proposal may be returned to the institution.

§ 46.104 [Reserved]

§ 46.105 [Reserved]

§ 46.106 [Reserved]

§ 46.107 IRB membership.

(a) Each IRB shall have at least five members, with varying backgrounds to promote complete and adequate review of research activities commonly conducted by the institution. The IRB shall be sufficiently qualified through the experience and expertise of its members, and the diversity of the members' backgrounds including consideration of the racial and cultural backgrounds of members and sensitivity to such issues as community attitudes, to promote respect for its advice and counsel in safeguarding the rights and welfare of human subjects. In addition to possessing the professional competence necessary to review specific research activities, the IRB shall be able to ascertain the acceptability of proposed research in terms of institutional commitments and regulations, applicable law, and standards of professional conduct and practice. The IRB shall therefore include persons knowledgeable in these areas. If an IRB regularly reviews research that involves a vulnerable category of subjects, including but not limited to subjects covered by other subparts of this part, the IRB shall include one or more individuals who are primarily concerned with the welfare of these subjects.

(b) No IRB may consist entirely or men or entirely of women, or entirely of members of one profession.

(c) Each IRB shall include at least one member whose primary concerns are in nonscientific areas; for example: lawyers, ethicists, members of the clergy.

(d) Each IRB shall include at least one member who is not otherwise affiliated with the institution and who is not part of the immediate family of a person who is affiliated with the institution.

(e) No IRB may have a member participating in the IRB's initial or continuing review of any project in which the member has a conflicting interest, except to provide information requested by the IRB.

(f) An IRB may, in its discretion, invite individuals with competence in special areas to assist in the review of complex

issues which require expertise beyond or in addition to that available on the IRB. These individuals may not vote with the IRB.

§ 46.108 IRB functions and operations.

In order to fulfill the requirements of these regulations each IRB shall:

(a) Follow written procedures as provided in § 46.103(b)(4).

(b) Except when an expedited review procedure is used (see § 46.110), review proposed research at convened meetings at which a majority of the members of the IRB are present, including at least one member whose primary concerns are in nonscientific areas. In order for the research to be approved, it shall receive the approval of a majority of those members present at the meeting.

(c) Be responsible for reporting to the appropriate institutional officials and the Secretary any serious or continuing noncompliance by investigators with the requirements and determinations of the IRB.

§ 46.109 IRB review of research.

(a) An IRB shall review and have authority to approve, require modifications in (to secure approval), or disapprove all research activities covered by these regulations.

(b) An IRB shall require that information given to subjects as part of informed consent is in accordance with § 46.116. The IRB may require that information, in addition to that specifically mentioned in § 46.116, be given to the subjects when in the IRB's judgment the information would meaningfully add to the protection of the rights and welfare of subjects.

(c) An IRB shall require documentation of informed consent or may waive documentation in accordance with § 46.117.

(d) An IRB shall notify investigators and the institution in writing of its decision to approve or disapprove the proposed research activity, or of modifications required to secure IRB approval of the research activity. If the IRB decides to disapprove a research activity, it shall include in its written notification a statement of the reasons for its decision and give the investigator an opportunity to respond in person or in writing.

(e) An IRB shall conduct continuing review of research covered by these regulations at intervals appropriate to the degree of risk, but not less than once per year, and shall have authority to observe or have a third party observe the consent process and the research.

Federal Register / Vol. 46, No. 16 / Monday, January 26, 1981 / Rules and Regulations   8389

§ 46.110 Expedited review procedures for certain kinds of research involving no more than minimal risk, and for minor changes in approved research.

(a) The Secretary has established, and published in the Federal Register, a list of categories of research that may be reviewed by the IRB through an expedited review procedure. The list will be amended, as appropriate, through periodic republication in the Federal Register.

(b) An IRB may review some or all of the research appearing on the list through an expedited review procedure, if the research involves no more than minimal risk. The IRB may also use the expedited review procedure to review minor changes in previously approved research during the period for which approval is authorized. Under an expedited review procedure, the review may be carried out by the IRB chairperson or by one or more experienced reviewers designated by the chairperson from among members of the IRB. In reviewing the research, the reviewers may exercise all of the authorities of the IRB except that the reviewers may not disapprove the research. A research activity may be disapproved only after review in accordance with the non-expedited procedure set forth in § 46.108(b).

(c) Each IRB which uses an expedited review procedure shall adopt a method for keeping all members advised of research proposals which have been approved under the procedure.

(d) The Secretary may restrict, suspend, or terminate an institution's or IRB's use of the expedited review procedure when necessary to protect the rights or welfare of subjects.

§ 46.111 Criteria for IRB approval of research.

(a) In order to approve research covered by these regulations the IRB shall determine that all of the following requirements are satisfied:

(1) Risks to subjects are minimized: (i) By using procedures which are consistent with sound research design and which do not unnecessarily expose subjects to risk, and (ii) whenever appropriate, by using procedures already being performed on the subjects for diagnostic or treatment purposes.

(2) Risks to subjects are reasonable in relation to anticipated benefits, if any, to subjects, and the importance of the knowledge that may reasonably be expected to result. In evaluating risks and benefits, the IRB should consider only those risks and benefits that may result from the research (as distinguished from risks and benefits of therapies subjects would receive even if

not participating in the research). The IRB should not consider possible long-range effects of applying knowledge gained in the research (for example, the possible effects of the research on public policy) as among those research risks that fall within the purview of its responsibility.

(3) Selection of subjects is equitable. In making this assessment the IRB should take into account the purposes of the research and the setting in which the research will be conducted.

(4) Informed consent will be sought from each prospective subject or the subject's legally authorized representative, in accordance with, and to the extent required by § 46.116.

(5) Informed consent will be appropriately documented, in accordance with, and to the extent required by § 46.117.

(6) Where appropriate, the research plan makes adequate provision for monitoring the data collected to insure the safety of subjects.

(7) Where appropriate, there are adequate provisions to protect the privacy of subjects and to maintain the confidentiality of data.

(b) Where some or all of the subjects are likely to be vulnerable to coercion or undue influence, such as persons with acute or severe physical or mental illness, or persons who are economically or educationally disadvantaged, appropriate additional safeguards have been included in the study to protect the rights and welfare of these subjects.

§ 46.112 Review by institution.

Research covered by these regulations that has been approved by an IRB may be subject to further appropriate review and approval or disapproval by officials of the institution. However, those officials may not approve the research if it has not been approved by an IRB.

§ 46.113 Suspension or termination of IRB approval of research.

An IRB shall have authority to suspend or terminate approval of research that is not being conducted in accordance with the IRB's requirements or that has been associated with unexpected serious harm to subjects. Any suspension or termination of approval shall include a statement of the reasons for the IRB's action and shall be reported promptly to the investigator, appropriate institutional officials, and the Secretary.

§ 46.114 Cooperative research.

Cooperative research projects are those projects, normally supported through grants, contracts, or similar arrangements, which involve institutions

in addition to the grantee or prime contractor (such as a contractor with the grantee, or a subcontractor with the prime contractor). In such instances, the grantee or prime contractor remains responsible to the Department for safeguarding the rights and welfare of human subjects. Also, when cooperating institutions conduct some or all of the research involving some or all of these subjects, each cooperating institution shall comply with these regulations as though it received funds for its participation in the project directly from the Department, except that in complying with these regulations institutions may use joint review, reliance upon the review of another qualified IRB, or similar arrangements aimed at avoidance of duplication of effort.

§ 46.115 IRB records.

(a) An institution, or where appropriate an IRB, shall prepare and maintain adequate documentation of IRB activities, including the following:

(1) Copies of all research proposals reviewed, scientific evaluations, if any, that accompany the proposals, approved sample consent documents, progress reports submitted by investigators, and reports of injuries to subjects.

(2) Minutes of IRB meetings which shall be in sufficient detail to show attendance at the meetings; actions taken by the IRB; the vote on these actions including the number of members voting for, against, and abstaining; the basis for requiring changes in or disapproving research; and a written summary of the discussion of controverted issues and their resolution.

(3) Records of continuing review activities.

(4) Copies of all correspondence between the IRB and the investigators.

(5) A list of IRB members as required by § 46.103(b)(3).

(6) Written procedures for the IRB as required by § 46.103(b)(4).

(7) Statements of significant new findings provided to subjects, as required by § 46.116(b)(5).

(b) The records required by this regulation shall be retained for at least 3 years after completion of the research, and the records shall be accessible for inspection and copying by authorized representatives of the Department at reasonable times and in a reasonable manner.

§ 46.116 General requirements for informed consent.

Except as provided elsewhere in this or other subparts, no investigator may involve a human being as a subject in

8390   Federal Register / Vol. 46, No. 16 / Monday, January 26, 1981 / Rules and Regulations

research covered by these regulations unless the investigator has obtained the legally effective informed consent of the subject or the subject's legally authorized representative. An investigator shall seek such consent only under circumstances that provide the prospective subject or the representative sufficient opportunity to consider whether or not to participate and that minimize the possibility of coercion or undue influence. The information that is given to the subject or the representative shall be in language understandable to the subject or the representative. No informed consent, whether oral or written, may include any exculpatory language through which the subject or the representative is made to waive or appear to waive any of the subject's legal rights, or releases or appears to release the investigator, the sponsor, the institution or its agents from liability for negligence.

(a) Basic elements of informed consent. Except as provided in paragraph (c) of this section, in seeking informed consent the following information shall be provided to each subject:

(1) A statement that the study involves research, an explanation of the purposes of the research and the expected duration of the subject's participation, a description of the procedures to be followed, and identification of any procedures which are experimental;

(2) A description of any reasonably foreseeable risks or discomforts to the subject;

(3) A description of any benefits to the subject or to others which may reasonably be expected from the research;

(4) A disclosure of appropriate alternative procedures or courses of treatment, if any, that might be advantageous to the subject;

(5) A statement describing the extent, if any, to which confidentiality of records identifying the subject will be maintained;

(6) For research involving more than minimal risk, an explanation as to whether any compensation and an explanation as to whether any medical treatments are available if injury occurs and, if so, what they consist of, or where further information may be obtained;

(7) An explanation of whom to contact for answers to pertinent questions about the research and research subjects' rights, and whom to contact in the event of a research-related injury to the subject; and

(8) A statement that participation is voluntary, refusal to participate will involve no penalty or loss of benefits to which the subject is otherwise entitled, and the subject may discontinue participation at any time without penalty or loss of benefits to which the subject is otherwise entitled.

(b) Additional elements of informed consent. When appropriate, one or more of the following elements of information shall also be provided to each subject:

(1) A statement that the particular treatment or procedure may involve risks to the subject (or to the embryo or fetus, if the subject is or may become pregnant) which are currently unforeseeable;

(2) Anticipated circumstances under which the subject's participation may be terminated by the investigator without regard to the subject's consent;

(3) Any additional costs to the subject that may result from participation in the research;

(4) The consequences of a subject's decision to withdraw from the research and procedures for orderly termination of participation by the subject;

(5) A statement that significant new findings developed during the course of the research which may relate to the subject's willingness to continue participation will be provided to the subject; and

(6) The approximate number of subjects involved in the study.

(c) An IRB may approve a consent procedure which does not include, or which alters, some or all of the elements of informed consent set forth above, or waive the requirement to obtain informed consent provided the IRB finds and documents that:

(1) The research is to be conducted for the purpose of demonstrating or evaluating: (i) Federal, state, or local benefit or service programs which are not themselves research programs, (ii) procedures for obtaining benefits or services under these programs, or (iii) possible changes in or alternatives to these programs or procedures; and

(2) The research could not practicably be carried out without the waiver or alteration.

(d) An IRB may approve a consent procedure which does not include, or which alters, some or all of the elements of informed consent set forth above, or waive the requirements to obtain informed consent provided the IRB finds and documents that:

(1) The research involves no more than minimal risk to the subjects;

(2) The waiver or alteration will not adversely affect the rights and welfare of the subjects;

(3) The research could not practicably be carried out without the waiver or alteration; and

(4) Whenever appropriate, the subjects will be provided with additional pertinent information after participation.

(e) The informed consent requirements in these regulations are not intended to preempt any applicable federal, state, or local laws which require additional information to be disclosed in order for informed consent to be legally effective.

(f) Nothing in these regulations is intended to limit the authority of a physician to provide emergency medical care, to the extent the physician is permitted to do so under applicable federal, state, or local law.

§ 46.117 Documentation of informed consent.

(a) Except as provided in paragraph (c) of this section, informed consent shall be documented by the use of a written consent form approved by the IRB and signed by the subject or the subject's legally authorized representative. A copy shall be given to the person signing the form.

(b) Except as provided in paragraph (c) of this section, the consent form may be either of the following:

(1) A written consent document that embodies the elements of informed consent required by § 46.116. This form may be read to the subject or the subject's legally authorized representative, but in any event, the investigator shall give either the subject or the representative adequate opportunity to read it before it is signed; or

(2) A "short form" written consent document stating that the elements of informed consent required by § 46.116 have been presented orally to the subject or the subject's legally authorized representative. When this method is used, there shall be a witness to the oral presentation. Also, the IRB shall approve a written summary of what is to be said to the subject or the representative. Only the short form itself is to be signed by the subject or the representative. However, the witness shall sign both the short form and a copy of the summary, and the person actually obtaining consent shall sign a copy of the summary. A copy of the summary shall be given to the subject or the representative, in addition to a copy of the "short form."

(c) An IRB may waive the requirement for the investigator to obtain a signed consent form for some or all subjects if it finds either:

(1) That the only record linking the subject and the research would be the consent document and the principal risk would be potential harm resulting from

Federal Register / Vol. 46, No. 16 / Monday, January 26, 1981 / Rules and Regulations   8391

a breach of confidentiality. Each subject will be asked whether the subject wants documentation linking the subject with the research, and the subject's wishes will govern; or

(2) That the research presents no more than minimal risk of harm to subjects and involves no procedures for which written consent is normally required outside of the research context.

In cases where the documentation requirement is waived, the IRB may require the investigator to provide subjects with a written statement regarding the research.

§ 46.118 Applications and proposals lacking definite plans for involvement of human subjects.

Certain types of applications for grants, cooperative agreements, or contracts are submitted to the Department with the knowledge that subjects may be involved within the period of funding, but definite plans would not normally be set forth in the application or proposal. These include activities such as institutional type grants (including bloc grants) where selection of specific projects is the institution's responsibility; research training grants where the activities involving subjects remain to be selected; and projects in which human subjects' involvement will depend upon completion of instruments, prior animal studies, or purification of compounds. These applications need not be reviewed by an IRB before an award may be made. However, except for research described in § 46.101(b). no human subjects may be involved in any project supported by these awards until the project has been reviewed and approved by the IRB, as provided in these regulations, and certification submitted to the Department.

§ 46.119 Research undertaken without the intention of involving human subjects.

In the event research (conducted or funded by the Department) is undertaken without the intention of involving human subjects, but it is later proposed to use human subjects in the research, the research shall first be reviewed and approved by an IRB, as provided in these regulations, a certification submitted to the Department, and final approval given to the proposed change by the Department.

46.120 Evaluation and disposition of applications and proposals.

(a) The Secretary will evaluate all applications and proposals involving human subjects submitted to the Department through such officers and employees of the Department and such experts and consultants as the Secretary

determines to be appropriate. This evaluation will take into consideration the risks to the subjects, the adequacy of protection against these risks, the potential benefits of the proposed research to the subjects and others, and the importance of the knowledge to be gained.

(b) On the basis of this evaluation, the Secretary may approve or disapprove the application or proposal, or enter into negotiations to develop an approvable one.

§ 46.121 Investigational new drug or device 30-day delay requirement.

When an institution is required to prepare or to submit a certification with an application or proposal under these regulations, and the application or proposal involves an investigational new drug (within the meaning of 21 U.S.C. 355(i) or 357(d)) or a significant risk device (as defined in 21 CFR 812.3(m)), the institution shall identify the drug or device in the certification. The institution shall also state whether the 30-day interval required for investigational new drugs by 21 CFR 312.1(a) and for significant risk devices by 21 CFR 812.30 has elapsed, or whether the Food and Drug Administration has waived that requirement. If the 30-day interval has expired, the institution shall state whether the Food and Drug Administration has requested that the sponsor continue to withhold or restrict the use of the drug or device in human subjects. If the 30-day interval has not expired, and a waiver has not been received, the institution shall send a statement to the Department upon expiration of the interval. The Department will not consider a certification acceptable until the institution has submitted a statement that the 30-day interval has elapsed, and the Food and Drug Administration has not requested it to limit the use of the drug or device, or that the Food and Drug Administration has waived the 30-day interval.

§ 46.122 Use of Federal funds.

Federal funds administered by the Department may not be expended for research involving human subjects unless the requirements of these regulations, including all subparts of these regulations, have been satisfied.

§ 46.123 Early termination of research funding; evaluation of subsequent applications and proposals.

(a) The Secretary may require that Department funding for any project be terminated or suspended in the manner prescribed in applicable program

requirements, when the Secretary finds an institution has materially failed to comply with the terms of these regulations.

(b) In making decisions about funding applications or proposals covered by these regulations the Secretary may take into account, in addition to all other eligibility requirements and program criteria, factors such as whether the applicant has been subject to a termination or suspension under paragraph (a) of this section and whether the applicant or the person who would direct the scientific and technical aspects of an activity has in the judgment of the Secretary materially failed to discharge responsibility for the protection of the rights and welfare of human subjects (whether or not Department funds were involved).

§ 46.124 Conditions.

With respect to any research project or any class of research projects the Secretary may impose additional conditions prior to or at the time of funding when in the Secretary's judgment additional conditions are necessary for the protection of human subjects.

[FR Doc. 81-4579 Filed 1-23-81 8.45 am]

BILLING CODE 4110-08-M

8392   Federal Register / Vol. 46, No. 16 / Monday, January 26, 1981 / Notices

DEPARTMENT OF HEALTH AND HUMAN SERVICES

Public Health Service

Research Activities Which May Be Reviewed Through Expedited Review Procedures Set Forth In HHS Regulations for Protection of Human Research Subjects

AGENCY: Department of Health and Human Services.

ACTION: Notice.

SUMMARY: This notice contains a list of research activities which Institutional Review Boards may review through the expedited review procedures set forth in HHS regulations for the protection of human subjects.

EFFECTIVE DATE: This Notice shall become effective on July 27, 1981. Institutions currently conducting or supporting research in accord with General Assurances negotiated with the Department of Health and Human Services (formerly HEW) may continue to do so in accord with conditions of their General Assurance. However these Institutions are permitted and encouraged to apply § 46.110 and the list of research categories, as soon as feasible. They need not wait for the effective date or the negotiation of a new assurance to operate under the new sections cited above. Institutions conducting or supporting research in accord with a Special Assurance negotiated with the Department, shall continue to do so until such time as the assurance terminates.

FOR FURTHER INFORMATION CONTACT: F. William Dommel, Jr., J.D., Assistant Director, Office for Protection from Research Risks, National Institutes of Health, 5333 Westbard Avenue, Room 3A18, Bethesda, Maryland 20205, telephone: (301) 496–7163.

SUPPLEMENTARY INFORMATION: Elsewhere in this issue of the Federal Register the Secretary is publishing final regulations relating to the protection of human subjects in research. The regulations amend Subpart A of 45 CFR Part 46.

Section 46.110 of the new final regulations provides that: "The Secretary will publish in the Federal Register a list of categories of research activities, involving no more than minimal risk, that may be reviewed by the Institutional Review Board, through an expedited review procedure * * *." This notice is published in accordance with § 46.110.

Research activities involving no more than minimal risk and in which the only involvement of human subjects will be

in one or more of the following categories (carried out through standard methods) may be reviewed by the Institutional Review Board through the expedited review procedure authorized in § 46.110 of 45 CFR Part 46.

(1) Collection of: hair and nail clippings, in a nondisfiguring manner; deciduous teeth; and permanent teeth if patient care indicates a need for extraction.

(2) Collection of excreta and external secretions including sweat, uncannulated saliva, placenta removed at delivery, and amniotic fluid at the time of rupture of the membrane prior to or during labor.

(3) Recording of data from subjects 16 years of age or older using noninvasive procedures routinely employed in clinical practice. This includes the use of physical sensors that are applied either to the surface of the body or at a distance and do not involve input of matter or significant amounts of energy into the subject or an invasion of the subject's privacy. It also includes such procedures as weighing, testing sensory acuity, electrocardiography, electroencephalography, thermography, detection of naturally occurring radioactivity, diagnostic echography, and electroretinography. It does not include exposure to electromagnetic radiation outside the visible range (for example, x-rays, microwaves).

(4) Collection of blood samples by venipuncture, in amounts not exceeding 450 milliliters in an eight-week period and no more often than two times per week, from subjects 18 years of age or older who are in good health and not pregnant.

(5) Collection of both supra- and subgingival dental plaque and calculus, provided the procedure is not more invasive than routine prophylactic scaling of the teeth and the process is accomplished in accordance with accepted prophylactic techniques.

(6) Voice recordings made for research purposes such as investigations of speech defects.

(7) Moderate exercise by healthy volunteers.

(8) The study of existing data, documents, records, pathological specimens, or diagnostic specimens.

(9) Research on individual or group behavior or characteristics of individuals, such as studies of perception, cognition, game theory, or test development, where the investigator does not manipulate subjects' behavior and the research will not involve stress to subjects.

(10) Research on drugs or devices for which an investigational new drug

exemption or an investigational device exemption is not required.

Dated: January 14, 1981.

Julius B. Richmond,
Assistant Secretary for Health and Surgeon General.

[FR Doc. 81–2506 Filed 1–23–81; 8.45 am]

BILLING CODE 4110–08–M

242

8950   **Federal Register** / Vol. 46, No. 17 / Tuesday, January 27, 1981 / Rules and Regulations

more of its own employees to conduct a clinical investigation it has initiated is considered to be a sponsor (not a sponsor-investigator), and the employees are considered to be investigators.

(f) "Sponsor-investigator" means an individual who both initiates and actually conducts, alone or with others, a clinical investigation, i.e., under whose immediate direction the test article is administered or dispensed to, or used involving, a subject. The term does not include any person other than an individual, e.g., corporation or agency.

(g) "Human subject" means an individual who is or becomes a participant in research, either as a recipient of the test article or as a control. A subject may be either a healthy human or a patient.

(h) "Institution" means any public or private entity or agency (including Federal, State, and other agencies). The word "facility" as used in section 520(g) of the act is deemed to be synonymous with the term "institution" for purposes of this part.

(i) "Institutional review board" (IRB) means any board, committee, or other group formally designated by an institution to review biomedical research involving humans as subjects, to approve the initiation of and conduct periodic review of such research. The term has the same meaning as the phrase "institutional review committee" as used in section 520(g) of the act.

(j) "Prisoner" means any individual involuntarily confined or detained in a penal institution. The term is intended to encompass individuals sentenced to such an institution under a criminal or civil statute, individuals detained in other facilities by virtue of statutes or commitment procedures that provide alternatives to criminal prosecution or incarceration in a penal institution, and individuals detained pending arraignment, trial, or sentencing.

(k) "Test article" means any drug (including a biological product for human use), medical device for human use, human food additive, color additive, electronic product, or any other article subject to regulation under the act or under sections 351 and 354–360F of the Public Health Service Act (42 U.S.C. 262 and 263b–263n).

(l) "Minimal risk" means that the risks of harm anticipated in the proposed research are not greater, considering probability and magnitude, than those ordinarily encountered in daily life or during the performance of routine physical or psychological examinations or tests.

(m) "Legally authorized representative" means an individual or

SUBCHAPTER A—GENERAL

PART 50—PROTECTION OF HUMAN SUBJECTS

1. In Part 50:
a. In § 50.3 by adding paragraphs (a) and (c) through (m), to read as follows:

§ 50.3   Definitions.

*   *   *   *   *

(a) "Act" means the Federal Food, Drug, and Cosmetic Act, as amended (secs. 201–902, 52 Stat. 1040 et seq. as amended (21 U.S.C. 321–392)).

*   *   *   *   *

(c) "Clinical investigation" means any experiment that involves a test article and one or more human subjects and that either is subject to requirements for prior submission to the Food and Drug Administration under section 505(i), 507(d), or 520(g) of the act, or is not subject to requirements for prior submission to the Food and Drug Administration under these sections of the act, but the results of which are intended to be submitted later to, or held for inspection by, the Food and Drug Administration as part of an application for a research or marketing permit. The term does not include experiments that are subject to the provisions of Part 58 of this chapter, regarding nonclinical laboratory studies.

(d) "Investigator" means an individual who actually conducts a clinical investigation, i.e., under whose immediate direction the test article is administered or dispensed to, or used involving, a subject, or, in the event of an investigation conducted by a team of individuals, is the responsible leader of that team.

(e) "Sponsor" means a person who initiates a clinical investigation, but who does not actually conduct the investigation, i.e., the test article is administered or dispensed to or used involving, a subject under the immediate direction of another individual. A person other than an individual (e.g., corporation or agency) that uses one or

Federal Register / Vol. 46, No. 17 / Tuesday, January 27, 1981 / Rules and Regulations   8951

judicial or other body authorized under applicable law to consent on behalf of a prospective subject to the subject's particpation in the procedure(s) involved in the research.

b. By adding new Subpart B to read as follows:

Subpart B—Informed Consent of Human Subjects

Sec.
50.20 General requirements for informed consent.
50.21 Effective date.
50.23 Exception from general requirements.
50.25 Elements of informed consent.
50.27 Documention of informed consent.

Subpart B—Informed Consent of Human Subjects

§ 50.20 General requirements for informed consent.

Except as provided in § 50.23, no investigator may involve a human being as a subject in research covered by these regulations unless the investigator has obtained the legally effective informed consent of the subject or the subject's legally authorized representative. An investigator shall seek such consent only under circumstances that provide the prospective subject or the representative sufficient opportunity to consider whether or not to participate and that minimize the possibility of coercion or undue influence. The information that is given to the subject or the representative shall be in language understandable to the subject or the representative. No informed consent, whether oral or written, may include any exculpatory language through which the subject or the representative is made to waive or appear to waive any of the subject's legal rights, or releases or appears to release the investigator, the sponsor, the institution, or its agents from liability for negligence.

§ 50.21 Effective date.

The requirements for informed consent set out in this part apply to all human subjects entering a clinical investigation that commences on or after July 1, 1981.

§ 50.23 Exception from general requirements.

(a) The obtaining of informed consent shall be deemed feasible unless, before use of the test article (except as provided in paragraph (b) of this section), both the investigator and a physician who is not otherwise participating in the clinical investigation certify in writing all of the following:
(1) The human subject is confronted by a life-threatening situation necessitating the use of the test article.

(2) Informed consent cannot be obtained from the subject because of an inability to communicate with, or obtain legally effective consent from, the subject.
(3) Time is not sufficient to obtain consent from the subject's legal representative.
(4) There is available no alternative method of approved or generally recognized therapy that provides an equal or greater likelihood of saving the life of the subject.
(b) If immediate use of the test article is, in the investigator's opinion, required to preserve the life of the subject, and time is not sufficient to obtain the independent determination required in paragraph (a) of this section in advance of using the test article, the determinations of the clinical investigator shall be made and, within 5 working days after the use of the article, be reviewed and evaluated in writing by a physician who is not participating in the clinical investigation.
(c) The documentation required in paragraph (a) or (b) of this section shall be submitted to the IRB within 5 working days after the use of the test article.

§ 50.25 Elements of informed consent.

(a) *Basic elements of informed consent.* In seeking informed consent, the following information shall be provided to each subject:
(1) A statement that the study involves research, an explanation of the purposes of the research and the expected duration of the subject's participation, a description of the procedures to be followed, and identification of any procedures which are experimental.
(2) A description of any reasonably foreseeable risks or discomforts to the subject.
(3) A description of any benefits to the subject or to others which may reasonably be expected from the research.
(4) A disclosure of appropriate alternative procedures or courses of treatment, if any, that might be advantageous to the subject.
(5) A statement describing the extent, if any, to which confidentiality of records identifying the subject will be maintained and that notes the possibility that the Food and Drug Administration may inspect the records.
(6) For research involving more than minimal risk, an explanation as to whether any compensation and an explanation as to whether any medical treatments are available if injury occurs and, if so, what they consist of, or where further information may be obtained.

(7) An explanation of whom to contact for answers to pertinent questions about the research and research subjects' rights, and whom to contact in the event of a research-related injury to the subject.
(8) A statement that participation is voluntary, that refusal to participate will involve no penalty or loss of benefits to which the subject is otherwise entitled, and that the subject may discontinue participation at any time without penalty or loss of benefits to which the subject is otherwise entitled.
(b) *Additional elements of informed consent.* When appropriate, one or more of the following elements of information shall also be provided to each subject:
(1) A statement that the particular treatment or procedure may involve risks to the subject (or to the embryo or fetus, if the subject is or may become pregnant) which are currently unforeseeable.
(2) Anticipated circumstances under which the subject's participation may be terminated by the investigator without regard to the subject's consent.
(3) Any additional costs to the subject that may result from participation in the research.
(4) The consequences of a subject's decision to withdraw from the research and procedures for orderly termination of participation by the subject.
(5) A statement that significant new findings developed during the course of the research which may relate to the subject's willingness to continue participation will be provided to the subject.
(6) The approximate number of subjects involved in the study.
(c) The informed consent requirements in these regulations are not intended to preempt any applicable Federal, State, or local laws which require additional information to be disclosed for informed consent to be legally effective.
(d) Nothing in these regulations is intended to limit the authority of a physician to provide emergency medical care to the extent the physician is permitted to do so under applicable Federal, State, or local law.

§ 50.27 Documentation of informed consent.

(a) Except as provided in § 56.109(c), informed consent shall be documented by the use of a written consent form approved by the IRB and signed by the subject or the subject's legally authorized representative. A copy shall be given to the person signing the form.
(b) Except as provided in § 56.109(c), the consent form may be either of the following:

(1) A written consent document that embodies the elements of informed consent required by § 50.25. This form may be read to the subject or the subject's legally authorized representative, but, in any event, the investigator shall give either the subject or the representative adequate opportunity to read it before it is signed.

(2) A "short form" written consent document stating that the elements of informed consent required by § 50.25 have been presented orally to the subject or the subject's legally authorized representative. When this method is used, there shall be a witness to the oral presentation. Also, the IRB shall approve a written summary of what is to be said to the subject or the representative. Only the short form itself is to be signed by the subject or the representative. However, the witness shall sign both the short form and a copy of the summary, and the person actually obtaining the consent shall sign a copy of the summary. A copy of the summary shall be given to the subject or the representative in addition to a copy of the short form.

PART 56—INSTITUTIONAL REVIEW BOARDS

Subpart A—General

Sec.
56.101   Scope.
56.102   Definitions.
56.103   Circumstances in which IRB review is required.
56.104   Exemptions from IRB requirement
56.105   Waiver of IRB requirement.

Subpart B—Organization and Personnel

56.107   IRB membership.

Subpart C—IRB Functions and Operations

56.108   IRB functions and operations
56.109   IRB review of research.
56.110   Expedited review procedures for certain kinds of research involving no more than minimal risk. and for minor changes in approved research
56.111   Criteria for IRB approval of research
56.112   Review by institution
56.113   Suspension or termination of IRB approval of research.

Sec.
56.114   Cooperative research

Subpart D—Records and Reports

56.115   IRB records

Subpart E—Administrative Action for Noncompliance

56.120   Lesser administrative actions
56.121   Disqualification of an IRB or an institution
56.122   Public disclosure of information regarding revocation
56.123   Reinstatement of an IRB or an institution
56.124   Actions alternative or additional to disqualification

Authority: Secs 406, 408, 409, 501, 502, 503, 505, 506, 507, 510, 513–516. 518–520, 701(a). 706. and 801. Pub. L. 717, 52 Stat 1049–1054 as amended, 1055, 1058 as amended. 55 Stat 851 as amended. 59 Stat 463 as amended. 68 Stat 511–518 as amended. 72 Stat 1785–1788 as amended, 74 Stat 399–407 as amended. 78 Stat 794–795 as amended. 90 Stat 540–546, 560, 562–574 (21 U S C 346, 346a. 348, 351, 352. 353, 355, 356, 357, 360. 360c–360f. 360h–360j. 371(a), 376, and 381). secs 215, 301, 351. 354–360f, Pub L. 410. 58 Stat 690. 702 as amended 82 Stat 1173–1186 as amended (42 U S C 216, 241, 262, 263b–263n).

Subpart A—General Provisions

§ 56.101 Scope.

(a) This part contains the general standards for the composition, operation, and responsibility of an Institutional Review Board (IRB) that reviews clinical investigations regulated by the Food and Drug Administration under sections 505(i), 507(d), and 520(g) of the act, as well as clinical investigations that support applications for research or marketing permits for products regulated by the Food and Drug Administration, including food and color additives, drugs for human use, medical devices for human use, biological products for human use, and electronic products. Compliance with this part is intended to protect the rights and welfare of human subjects involved in such investigations.

(b) References in this part to regulatory sections of the Code of Federal Regulations are to Chapter I of Title 21, unless otherwise noted.

§ 56.102 Definitions.

As used in this part.
(a) "Act" means the Federal Food, Drug, and Cosmetic Act, as amended (secs. 201–902, 52 Stat. 1040 et seq., as amended (21 U.S.C. 321–392))
(b) "Application for research or marketing permit" includes:
(1) A color additive petition, described in Part 71.
(2) Data and information regarding a substance submitted as part of the procedures for establishing that a substance is generally recognized as

safe for a use which results or may reasonably be expected to result, directly or indirectly, in its becoming a component or otherwise affecting the characteristics of any food. described in § 170.35.
(3) A food additive petition, described in Part 171.
(4) Data and information regarding a food additive submitted as part of the procedures regarding food additives permitted to be used on an interim basis pending additional study, described in § 180.1.
(5) Data and information regarding a substance submitted as part of the procedures for establishing a tolerance for unavoidable contaminants in food and food-packaging materials. described in section 406 of the act
(6) A "Notice of Claimed Investigational Exemption for a New Drug" described in Part 312.
(7) A new drug application, described in Part 314.
(8) Data and information regarding the bioavailability or bioequivalence of drugs for human use submitted as part of the procedures for issuing, amending, or repealing a bioequivalence requirement, described in Part 320
(9) Data and information regarding an over-the-counter drug for human use submitted as part of the procedures for classifying such drugs as generally recognized as safe and effective and not misbranded, described in Part 330.
(10) Data and information regarding an antibiotic drug submitted as part of the procedures for issuing, amending, or repealing regulations for such drugs. described in Part 430.
(11) An application for a biological product license, described in Part 601.
(12) Data and information regarding a biological product submitted as part of the procedures for determining that licensed biological products are safe and effective and not misbranded, as described in Part 601.
(13) An "Application for an Investigational Device Exemption," described in Parts 812 and 813.
(14) Data and information regarding a medical device for human use submitted as part of the procedures for classifying such devices, described in Part 860.
(15) Data and information regarding a medical device for human use submitted as part of the procedures for establishing, amending, or repealing a standard for such device. described in Part 861.
(16) An application for premarket approval of a medical device for human use, described in section 515 of the act
(17) A product development protocol for a medical device for human use, described in section 515 of the act.

8976   Federal Register / Vol. 46. No. 17 / Tuesday, January 27, 1981 / Rules and Regulations

(18) Data and information regarding an electronic product submitted as part of the procedures for establishing, amending, or repealing a standard for such products, described in section 358 of the Public Health Service Act.

(19) Data and information regarding an electronic product submitted as part of the procedures for obtaining a variance from any electronic product performance standard, as described in § 1010.4.

(20) Data and information regarding an electronic product submitted as part of the procedures for granting, amending, or extending an exemption from a radiation safety performance standard, as described in § 1010.5.

(21) Data and information regarding an electronic product submitted as part of the procedures for obtaining an exemption from notification of a radiation safety defect or failure of compliance with a radiation safety performance standard, described in Subpart D of Part 1003.

(c) "Clinical investigation" means any experiment that involves a test article and one or more human subjects, and that either must meet the requirements for prior submission to the Food and Drug Administration under section 505(i), 507(d), or 520(g) of the act, or need not meet the requirements for prior submission to the Food and Drug Administration under these sections of the act, but the results of which are intended to be later submitted to, or held for inspection by, the Food and Drug Administration as part of an application for a research or marketing permit. The term does not include experiments that must meet the provisions of Part 58, regarding nonclinical laboratory studies. The terms "research," "clinical research," "clinical study," "study," and "clinical investigation" are deemed to be synonymous for purposes of this part.

(d) "Emergency use" means the use of a test article on a human subject in a life-threatening situation in which no standard acceptable treatment is available, and in which there is not sufficient time to obtain IRB approval.

(e) "Human subject" means an individual who is or becomes a participant in research, either as a recipient of the test article or as a control. A subject may be either a healthy individual or a patient.

(f) "Institution" means any public or private entity or agency (including Federal, State, and other agencies). The term "facility" as used in section 520(g) of the act is deemed to be synonymous with the term "institution" for purposes of this part.

(g) "Institutional Review Board (IRB)" means any board, committee, or other group formally designated by an institution to review, to approve the initiation of, and to conduct periodic review of, biomedical research involving human subjects. The primary purpose of such review is to assure the protection of the rights and welfare of the human subjects. The term has the same meaning as the phrase "institutional review committee" as used in section 520(g) of the act.

(h) "Investigator" means an individual who actually conducts a clinical investigation (i.e., under whose immediate direction the test article is administered or dispensed to, or used involving, a subject) or, in the event of an investigation conducted by a team of individuals, is the responsible leader of that team.

(i) "Minimal risk" means that the risks of harm anticipated in the proposed research are not greater, considering probability and magnitude, than those ordinarily encountered in daily life or during the performance of routine physical or psychological examinations or tests.

(j) "Sponsor" means a person or other entity that initiates a clinical investigation, but that does not actually conduct the investigation, i.e., the test article is administered or dispensed to, or used involving, a subject under the immediate direction of another individual. A person other than an individual (e.g., a corporation or agency) that uses one or more of its own employees to conduct an investigation that it has initiated is considered to be a sponsor (not a sponsor-investigator), and the employees are considered to be investigators.

(k) "Sponsor-investigator" means an individual who both initiates and actually conducts, alone or with others, a clinical investigation, i.e., under whose immediate direction the test article is administered or dispensed to, or used involving, a subject. The term does not include any person other than an individual, e.g., it does not include a corporation or agency. The obligations of a sponsor-investigator under this part include both those of a sponsor and those of an investigator.

(l) "Test article" means any drug for human use, biological product for human use, medical device for human use, human food additive, color additive, electronic product, or any other article subject to regulation under the act or under sections 351 or 354–360F of the Public Health Service Act.

§ 56.103   Circumstances in which IRB review is required.

(a) Except as provided in §§ 56.104 and 56.105, any clinical investigation which must meet the requirements for prior submission (as required in Parts 312, 812, and 813) to the Food and Drug Administration shall not be initiated unless that investigation has been reviewed and approved by, and remains subject to continuing review by, an IRB meeting the requirements of this part. The determination that a clinical investigation of this part.

(b) Except as provided in §§ 56.104 and 56.105, the Food and Drug Administration may decide not to consider in support of an application for a research or marketing permit any data or information that has been derived from a clinical investigation that has not been approved by, and that was not subject to initial and continuing review by, an IRB meeting the requirements may not be considered in support of an application for a research or marketing permit does not, however, relieve the applicant for such a permit of any obligation under any other applicable regulations to submit the results of the investigation to the Food and Drug Administration.

(c) Compliance with these regulations will in no way render inapplicable pertinent Federal, State, or local laws or regulations.

§ 56.104   Exemptions from IRB requirement.

The following categories of clinical investigations are exempt from the requirements of this part for IRB review:

(a) Any investigation which commenced before July 27, 1981, and was subject to requirements for IRB review under FDA regulations before that date, provided that the investigation remains subject to review of an IRB which meets the FDA requirements in effect before July 27, 1981.

(b) Any investigation commenced before July 27, 1981, and was not otherwise subject to requirements for IRB review under Food and Drug Administration regulations before that date.

(c) Emergency use of a test article, provided that such emergency use is reported to the IRB within 5 working days. Any subsequent use of the test article at the institution is subject to IRB review.

§ 56.105   Waiver of IRB requirement.

On the application of a sponsor or sponsor-investigator, the Food and Drug Administration may waive any of the requirements contained in these regulations, including the requirements for IRB review, for specific research activities or for classes of research activities, otherwise covered by these regulations.

Subpart B—Organization and Personnel

§ 56.107   IRB membership.

(a) Each IRB shall have at least five members, with varying backgrounds to promote complete and adequate review of research activities commonly conducted by the institution. The IRB shall be sufficiently qualified through the experience and expertise of its members, and the diversity of the members' backgrounds including consideration of the racial and cultural backgrounds of members and sensitivity to such issues as community attitudes, to promote respect for its advice and counsel in safeguarding the rights and welfare of human subjects. In addition to possessing the professional competence necessary to review specific research activities, the IRB shall be able to ascertain the acceptability of proposed research in terms of institutional commitments and regulations, applicable law, and standards of professional conduct and practice. The IRB shall therefore include persons knowledgeable in these areas. If an IRB regularly reviews research that involves a vulnerable category of subjects, including but not limited to subjects covered by other parts of this chapter, the IRB should include one or more individuals who are primarily concerned with the welfare of these subjects.

(b) No IRB may consist entirely of men, or entirely of women, or entirely of members of one profession.

(c) Each IRB shall include at least one member whose primary concerns are in nonscientific areas; for example: lawyers, ethicists, members of the clergy.

(d) Each IRB shall include at least one member who is not otherwise affiliated with the institution and who is not part of the immediate family of a person who is affiliated with the institution.

(e) No IRB may have a member participate in the IRB's initial or continuing review of any project in which the member has a conflicting interest, except to provide information requested by the IRB.

(f) An IRB may, in its discretion, invite individuals with competence in special areas to assist in the review of complex issues which require expertise beyond or in addition to that available on the IRB. These individuals may not vote with the IRB.

Subpart C—IRB Functions and Operations

§ 56.108   IRB functions and operations.

In order to fulfill the requirements of these regulations, each IRB shall:

(a) Follow written procedures (1) for conducting its initial and continuing review of research and for reporting its findings and actions to the investigator and the institution, (2) for determining which projects require review more often than annually and which projects need verification from sources other than the investigators that no material changes have occurred since previous IRB review, (3) for insuring prompt reporting to the IRB of changes in a research activity, (4) for insuring that changes in approved research, during the period for which IRB approval has already been given, may not be initiated without IRB review and approval except where necessary to eliminate apparent immediate hazards to the human subjects; and (5) for insuring prompt reporting to the IRB of unanticipated problems involving risks to subjects or others.

(b) Except when an expedited review procedure is used (see § 56.110), review proposed research at convened meetings at which a majority of the members of the IRB are present, including at least one member whose primary concerns are in nonscientific areas. In order for the research to be approved, it shall receive the approval of a majority of those members present at the meeting.

(c) Be responsible for reporting to the appropriate institutional officials and the Food and Drug Administration any serious or continuing noncompliance by investigators with the requirements and determinations of the IRB.

§ 56.109   IRB review of research.

(a) An IRB shall review and have authority to approve, require modifications in (to secure approval), or disapprove all research activities covered by these regulations.

(b) An IRB shall require that information given to subjects as part of informed consent is in accordance with § 50.25. The IRB may require that information, in addition to that specifically mentioned in § 50.25, be given to the subjects when in the IRB's judgment the information would meaningfully add to the protection of the rights and welfare of subjects.

(c) An IRB shall require documentation of informed consent in accordance with § 50.27, except that the IRB may, for some or all subjects, waive the requirement that the subject or the subject's legally authorized representative sign a written consent

form if it finds that the research presents no more than minimal risk of harm to subjects and involves no procedures for which written consent is normally required outside the research context. In cases where the documentation requirement is waived, the IRB may require the investigator to provide subjects with a written statement regarding the research.

(d) An IRB shall notify investigators and the institution in writing of its decision to approve or disapprove the proposed research activity, or of modifications required to secure IRB approval of the research activity. If the IRB decides to disapprove a research activity, it shall include in its written notification a statement of the reasons for its decision and give the investigator an opportunity to respond in person or in writing.

(e) An IRB shall conduct continuing review of research covered by these regulations at intervals appropriate to the degree of risk, but not less than once per year, and shall have authority to observe or have a third party observe the consent process and the research.

§ 56.110   Expedited review procedures for certain kinds of research involving no more than minimal risk, and for minor changes in approved research.

(a) The Food and Drug Administration has established, and published in the Federal Register, a list of categories of research that may be reviewed by the IRB through an expedited review procedure. The list will be amended, as appropriate, through periodic republication in the Federal Register.

(b) An IRB may review some or all of the research appearing on the list through an expedited review procedure, if the research involves no more than minimal risk. The IRB may also use the expedited review procedure to review minor changes in previously approved research during the period for which approval is authorized. Under an expedited review procedure, the review may be carried out by the IRB chairperson or by one or more experienced reviewers designated by the chairperson from among members of the IRB. In reviewing the research, the reviewers may exercise all of the authorities of the IRB except that the reviewers may not disapprove the research. A research activity may be disapproved only after review in accordance with the non-expedited procedure set forth in § 56.108(b).

(c) Each IRB which uses an expedited review procedure shall adopt a method for keeping all members advised of research proposals which have been approved under the procedure.

8978   Federal Register / Vol. 46, No. 17 / Tuesday, January 27, 1981 / Rules and Regulations

(d) The Food and Drug Administration may restrict, suspend, or terminate an institution's or IRB's use of the expedited review procedure when necessary to protect the rights or welfare of subjects.

§ 56.111 Criteria for IRB approval of research.

(a) In order to approve research covered by these regulations the IRB shall determine that all of the following requirements are satisfied:

(1) Risks to subjects are minimized: (i) by using procedures which are consistent with sound research design and which do not unnecessarily expose subjects to risk, and (ii) whenever appropriate, by using procedures already being performed on the subjects for diagnostic or treatment purposes.

(2) Risks to subjects are reasonable in relation to anticipated benefits, if any, to subjects, and the importance of the knowledge that may be expected.to result. In evaluating risks and benefits, the IRB should consider only those risks and benefits that may result from the research (as distinguished from risks and benefits of therapies that subjects would receive even if not participating in the research). The IRB should not consider possible long-range effects of applying knowledge gained in the research (for example, the possible effects of the research on public policy) as among those research risks that fall within the purview of its responsibility.

(3) Selection of subjects is equitable. In making this assessment, the IRB should take into account the purposes of the research and the setting in which the research will be conducted.

(4) Informed consent will be sought from each prospective subject or the subject's legally authorized representative, in accordance with and to the extent required by Part 50.

(5) Informed consent will be appropriately documented, in accordance with and to the extent required by § 50.27.

(6) Where appropriate, the research plan makes adequate provision for monitoring the data collected to ensure the safety of subjects.

(7) Where appropriate, there are adequate provisions to protect the privacy of subjects and to maintain the confidentiality of data.

(b) Where some or all of the subjects are likely to be vulnerable to coercion or undue influence, such as persons with acute or severe physical or mental illness, or persons who are economically or educationally disadvantaged, appropriate additional safeguards have been included in the study to protect the rights and welfare of these subjects.

§ 56.112   Review by institution.

Research covered by these regulations that has been approved by an IRB may be subject to further appropriate review and approval or disapproval by officials of the institution. However, those officials may not approve the research if it has not been approved by an IRB.

§ 56.113   Suspension or termination of IRB approval of research.

An IRB shall have authority to suspend or terminate approval of research that is not being conducted in accordance with the IRB's requirements or that has been associated with unexpected serious harm to subjects. Any suspension or termination of approval shall include a statement of the reasons for the IRB's action and shall be reported promptly to the investigator, appropriate institutional officials, and the Food and Drug Administration.

§ 56.114   Cooperative research.

In complying with these regulations, institutions involved in multi-institutional studies may use joint review, reliance upon the review of another qualified IRB, or similar arrangements aimed at avoidance of duplication of effort.

Subpart D—Records and Reports

§ 56.115   IRB records.

(a) An institution, or where appropriate an IRB, shall prepare and maintain adequate documentation of IRB activities, including the following:

(1) Copies of all research proposals reviewed, scientific evaluations, if any, that accompany the proposals, approved sample consent documents, progress reports submitted by investigators, and reports of injuries to subjects.

(2) Minutes of IRB meetings which shall be in sufficient detail to show attendance.at the meetings; actions taken by the IRB; the vote on these actions including the number of members voting for, against, and abstaining; the basis for requiring changes in or disapproving research; and a written summary of the discussion of controverted issues and their resolution.

(3) Records of continuing review activities.

(4) Copies of all correspondence between the IRB and the investigators.

(5) A list of IRB members identified by name; earned degrees; representative capacity; indications of experience such as board certifications, licenses, etc., sufficient to describe each member's chief anticipated contributions to IRB deliberations; and any employment or other relationship between each

member and the institution; for example: full-time employee, part-time employee, a member of governing panel or board, stockholder, paid or unpaid consultant.

(6) Written procedures for the IRB as required by § 56.108(a).

(7) Statements of significant new findings provided to subjects, as required by § 50.25.

(b) The records required by this regulation shall be retained for at least 3 years after completion of the research, and the records shall be accessible for inspection and copying by authorized representatives of the Food and Drug Administration at reasonable times and in a reasonable manner.

(c) The Food and Drug Administration may refuse to consider a clinical investigation in support of an application for a research or marketing permit if the institution or the IRB that reviewed the investigation refuses to allow an inspection under this section.

Subpart E—Administrative Actions for Noncompliance

§ 56.120   Lesser administrative actions.

(a) If apparent noncompliance with these regulations in the operation of an IRB is observed by an FDA investigator during an inspection, the inspector will present an oral or written summary of observations to an appropriate representative of the IRB. The Food and Drug Administration may subsequently send a letter describing the noncompliance to the IRB and to the parent institution. The agency will require that the IRB or the parent institution respond to this letter within a time period specified by FDA and describe the corrective actions that will be taken by the IRB, the institution, or both to achieve compliance with these regulations.

(b) On the basis of the IRB's or the institution's response, FDA may schedule a reinspection to confirm the adequacy of corrective actions. In addition, until the IRB or the parent institution takes appropriate corrective action, the agency may:

(1) Withhold approval of new studies subject to the requirements of this part that are conducted at the institution or reviewed by the IRB;

(2) Direct that no new subjects be added to ongoing studies subject to this part;

(3) Terminate ongoing studies subject to this part when doing so would not endanger the subjects; or

(4) When the apparent noncompliance creates a significant threat to the rights and welfare of human subjects, notify relevant State and Federal regulatory agencies and other parties with a direct

248

interest in the agency's action of the deficiencies in the operation of the IRB.

(c) The parent institution is presumed to be responsible for the operation of an IRB, and the Food and Drug Administration will ordinarily direct any administrative action under this subpart against the institution. However, depending on the evidence of responsibility for deficiencies, determined during the investigation, the Food and Drug Administration may restrict its administrative actions to the IRB or to a component of the parent institution determined to be responsible for formal designation of the IRB.

§ 56.121 Disqualification of an IRB or an institution.

(a) Whenever the IRB or the institution has failed to take adequate steps to correct the noncompliance stated in the letter sent by the agency under § 56.120(a), and the Commissioner of Food and Drugs determines that this noncompliance may justify the disqualification of the IRB or of the parent institution, the Commissioner will institute proceedings in accordance with the requirements for a regulatory hearing set forth in Part 16.

(b) The Commissioner may disqualify an IRB or the parent institution if the Commissioner determines that:

(1) The IRB has refused or repeatedly failed to comply with any of the regulations set forth in this part, and

(2) The noncompliance adversely affects the rights or welfare of the human subjects in a clinical investigation.

(c) If the Commissioner determines that disqualification is appropriate, the Commissioner will issue an order that explains the basis for the determination and that prescribes any actions to be taken with regard to ongoing clinical research conducted under the review of the IRB. The Food and Drug Administration will send notice of the disqualification to the IRB and the parent institution. Other parties with a direct interest, such as sponsors and clinical investigators, may also be sent a notice of the disqualification. In addition, the agency may elect to publish a notice of its action in the **Federal Register.**

(d) The Food and Drug Administration will not approve an application for a research permit for a clinical investigation that is to be under the review of a disqualified IRB or that is to be conducted at a disqualified institution, and it may refuse to consider in support of a marketing permit the data from a clinical investigation that was reviewed by a disqualified IRB as conducted at a disqualified institution.

unless the IRB or the parent institution is reinstated as provided in § 56.123.

§ 56.122 Public disclosure of information regarding revocation.

A determination that the Food and Drug Administration has disqualified an institution and the administrative record regarding that determination are disclosable to the public under Part 20.

§ 56.123 Reinstatement of an IRB or an institution.

An IRB or an institution may be reinstated if the Commissioner determines, upon an evaluation of a written submission from the IRB or institution that explains the corrective action that the institution or IRB plans to take, that the IRB or institution has provided adequate assurance that it will operate in compliance with the standards set forth in this part. Notification of reinstatement shall be provided to all persons notified under § 56.121(c).

§ 56.124 Actions alternative or additional to disqualification.

Disqualification of an IRB or of an institution is independent of, and neither in lieu of nor a precondition to, other proceedings or actions authorized by the act. The Food and Drug Administration may, at any time, through the Department of Justice institute any appropriate judicial proceedings (civil or criminal) and any other appropriate regulatory action, in addition to or in lieu of, and before, at the time of, or after, disqualification. The agency may also refer pertinent matters to another Federal, State, or local government agency for any action that that agency determines to be appropriate.

Effective date. This regulation shall become effective July 27, 1981.

(Secs. 406, 408, 409, 501, 502, 503, 505, 506, 507, 510, 513–516, 518–520, 701(a), 706, and 801, 52 Stat. 1049–1054 as amended, 1055, 1058 as amended, 55 Stat. 851 as amended, 59 Stat. 463 as amended, 68 Stat. 511–517 as amended, 72 Stat. 1785–1788 as amended, 74 Stat. 399–407 as amended, 76 Stat. 794–795 as amended, 90 Stat. 540–560, 562–574 (21 U.S.C. 346, 346a, 348, 351, 352, 353, 355, 356, 357, 360, 360c–360f, 360h–360j, 371(a) 376, and 381); secs. 215, 301, 351, as amended (42 U.S.C. 216, 241, 262, 263b–263n))

Dated: January 19, 1981.

Jere E. Goyan,

Commissioner of Food and Drugs

[FR Doc 81–2666 Filed 1–26–81 8 45 am]

BILLING CODE 4110–03–M

8980   Federal Register / Vol. 46, No. 17 / Tuesday, January 27, 1981 / Notices

DEPARTMENT OF HEALTH AND HUMAN SERVICES

Food and Drug Administration

[Docket No. 77N-0350]

Protection of Human Research Subjects; Clinical Investigations Which May Be Reviewed Through Expedited Review Procedure Set forth in FDA Regulations

AGENCY: Food and Frug Administration.

ACTION: Notice.

SUMMARY: This notice contains a list of research activities which institutional review boards may review through the expedited review procedures set forth in FDA regulations for the protection of human research subjects.

FOR FURTHER INFORMATION CONTACT: John C. Petricciani, Office of the Commissioner (HFB-4), Food and Drug Administration, 8800 Rockville Pike, Bethesda, MD 20205, 301-496-9320.

SUPPLEMENTARY INFORMATION: Elsewhere in this issue of the Federal Register, the Food and Drug Administration (FDA) is publishing final regulations establishing standards for institutional review boards (IRBs) for clinical investigations relating to the protection of human subjects in research. Section 56.110 (21 CFR 56.110) of the final IRB regulations provides that the agency will publish in the Federal Register a list of categories of research activities, involving no more than minimal risk, that may be reviewed by an IRB through expedited review procedures. This notice is published in accordance with § 56.110.

The agency concludes that research activities with human subjects involving no more than minimal risk and involving one or more of the following categories (carried out through standard methods), may be reviewed by an IRB through the expedited review procedure authorized in § 56.110.

(1) Collection of hair and nail clippings in a non-disfiguring manner; of deciduous teeth; and of permanent teeth if patient care indicates a need for extraction.

(2) Collection of excreta and external secretions including sweat and uncannulated saliva; of placenta at delivery; and of amniotic fluid at the time of rupture of the membrane before or during labor.

(3) Recording of data from subjects who are 18 years of age or older using noninvasive procedures routinely employed in clinical practice. This category includes the use of physical sensors that are applied either to the surface of the body or at a distance and do not involve input of matter or significant amounts of energy into the subject or an invasion of the subject's privacy. It also includes such procedures as weighting, electrocardiography, electroencephalography, thermography, detection of naturally occurring radioactivity, diagnostic echography, and electroretinography. This category does not include exposure to electromagnetic radiation outside the visible range (for example, x-rays or microwaves).

(4) Collection of blood samples by venipuncture, in amounts not exceeding 450 milliliters in an eight-week period and no more often than two times per week, from subjects who are 18 years of age or older and who are in good health and not pregnant.

(5) Collection of both supra- and subgingival dental plaque and calculus, provided the procedure is not more invasive than routine prophylactic scaling of the teeth, and the process is accomplished in accordance with accepted prophylactic techniques.

(6) Voice recordings made for research purposes such as investigations of speech defects.

(7) Moderate exercise by healthy volunteers.

(8) The study of existing data, documents, records, pathological specimens, or diagnostic specimens.

(9) Research on drugs or devices for which an investigational new drug exemption or an investigational device exemption is not required.

This list will be amended as appropriate and a current list will be published periodically to the Federal Register.

Dated: January 19, 1981.

Jere E. Goyan,

Commissioner of Food and Drugs.

[FR Doc. 81-2800 Filed 1-21-81, 3.56 pm]

BILLING CODE 4110-03-M

PART 812—INVESTIGATIONAL DEVICE EXEMPTIONS

Subpart A—General Provisions

Sec.
812.1   Scope.
812.2   Applicability.
812.3   Definitions.
812.5   Labeling of investigational devices.
812.7   Prohibition of promotion and other practices.
812.10   Waivers.
812.18   Import and export requirements.
812.19   Address for correspondence.

Subpart B—Application and Administrative Action

812.20   Application.
812.25   Investigational plan.
812.27   Report of prior investigations.
812.30   FDA action on applications.
812.35   Supplemental applications.
812.36   Confidentiality of data and information.

Subpart C—Responsibilities of Sponsors

812.40   General responsibilities of sponsors.
812.43   Selecting investigators and monitors.
812.45   Informing investigators.
812.46   Monitoring investigations.

Subpart D—Responsibilities of Institutional Review Boards

812.60   General responsibilities of Institutional Review Boards.
812.62   Membership.
812.65   Procedures.
812.70   Review of IRB actions.

Subpart E—Responsibilities of Investigators.

812.100   General responsibilities of investigators.
812.110   Specific responsibilities of investigators.

Subpart F—Informed Consent

812.120   General.
812.122   Requirements.
812.123   Exception.
812.130   Elements of informed consent.

Subpart G—Records and Reports

812.140   Records.
812.145   Inspections.
812.150   Reports.

Subpart A—General Provisions

§ 812.1   Scope.

(a) The purpose of this part is to encourage, to the extent consistent with the protection of public health and safety and with ethical standards, the discovery and development of useful devices intended for human use, and to that end to maintain optimum freedom for scientific investigators in their pursuit of this purpose. This part provides procedures for the conduct of clinical investigations of devices. An approved investigational device exemption (IDE) permits a device that otherwise would be required to comply with a performance standard or to have premarket approval to be shipped lawfully for the purpose of conducting investigations of that device. An IDE approved under § 812.30 or considered approved under § 812.2(b) exempts a device from the requirements of the following sections of the act and regulations issued thereunder: Misbranding under section 502, registration, listing, and premarket notification under section 510, performance standards under section 514, premarket approval under section 515, a banned device regulation under section 516, records and reports under section 519, restricted device requirements under section 520(e), good manufacturing practice requirements under section 520(f) (unless the sponsor states an intention to comply with these requirements under § 812.20(b)(3) or § 812.140(b)(4)(v)) and color additive requirements under section 706.

(b) References in this part to regulatory sections of the Code of Federal Regulations are to Chapter I of Title 21, unless otherwise noted.

§ 812.2   Applicability.

(a) *General.* This part applies to all clinical investigations of devices to determine safety and effectiveness, except as provided in paragraph (c) of this section.

(b) *Abbreviated requirements.* The following categories of investigations are considered to have approved applications for IDE's, unless FDA has notified a sponsor under § 812.20(a) that approval of an application is required:

(1) An investigation of a device other than a significant risk device, if the device is not a banned device and the sponsor:

(i) Labels the device in accordance with § 812.5;

(ii) Obtains IRB approval of the investigation after presenting the reviewing IRB with a brief explanation of why the device is not a significant risk device, and maintains such approval;

(iii) Ensures that each investigator participating in an investigation of the device obtains and documents informed consent under Subpart F for each subject under the investigator's care;

(iv) Complies with the requirements of § 812.46 with respect to monitoring investigations;

(v) Maintains the records required under § 812.140(b) (4) and (5) and makes the reports required under § 812.150(b) (1) through (3) and (5) through (10);

(vi) Ensures that participating investigators maintain the records required by § 812.140(a)(3)(i) and make the reports required under § 812.150(a) (1), (2), (5), and (7); and

(vii) Complies with the prohibitions in § 812.7 against promotion and other practices.

(2) An investigation of a device other than one subject to paragraph (e) of this section, if the investigation was begun on or before July 16, 1980, and to be completed, and is completed, on or before January 19, 1981.

(c) *Exempted investigations.* This part does not apply to investigations of the following categories of devices:

(1) A device, other than a transitional device, in commercial distribution immediately before May 28, 1976, when used or investigated in accordance with the indications in labeling in effect at that time.

(2) A device, other than a transitional device, introduced into commercial distribution on or after May 28, 1976, that FDA has determined to be substantially equivalent to a device in commercial distribution immediately before May 28, 1976, and that is used or investigated in accordance with the indications in the labeling FDA reviewed under Subpart E of Part 807 in determining substantial equivalence.

(3) A diagnostic device, if the sponsor complies with applicable requirements in § 809.10(c) and if the testing (i) is noninvasive, (ii) does not require an invasive sampling procedure that presents significant risk, (iii) does not by design or intention introduce energy into a subject, and (iv) is not used as a diagnostic procedure without confirmation of the diagnosis by another, medically established diagnostic product or procedure.

(4) A device undergoing consumer preference testing, testing of a modification, or testing of a combination of two or more devices in commercial distribution, if the testing is not for the purpose of determining safety or effectiveness and does not put subjects at risk.

(5) A device intended solely for veterinary use.

(6) A device shiped solely for research on or with laboratory animals and labeled in accordance with § 812.5(c).

(7) A custom device as defined in § 812.3(b), unless the device is being used to determine safety or effectiveness for commercial distribution.

(8) An intraocular lens. An intraocular lens shall not be used unless it is subject to an approved IDE under Part 813 or an approved application for premarket approval under section 515 of the act.

(d) *Limit on certain exemptions.* In the case of class II or class III device described in paragraph (c)(1) or (2) of this section, this part applies beginning on the date stipulated in an FDA

3752   Federal Register / Vol. 45, No. 13 / Friday, January 18, 1980 / Rules and Regulations

regulation or order that calls for the submission of premarket approval applications for an unapproved class III device, or establishes a performance standard for a class II device.

(e) *Investigations subject to IND's.* A sponsor that, on July 16, 1980, has an effective investigational new drug exemption (IND) for an investigation of a device shall continue to comply with the requirements of Part 312 until 90 days after that date. To continue the investigation after that date, a sponsor shall comply with paragraph (b)(1) of this section, if the device is not a significant risk device, or shall have obtained FDA approval under § 812.30 of an IDE application for the investigation of the device.

§ 812.3 Definitions.

(a) "Act" means the Federal Food, Drug, and Cosmetic Act (sections 201–901, 52 Stat. 1040 et seq., as amended (21 U.S.C. 301–392)).

(b) "Custom device" means a device that:

(1) Necessarily deviates from devices generally available or from an applicable performance standard or premarket approval requirement in order to comply with the order of an individual physician or dentist;

(2) Is not generally available to, or generally used by, other physicians or dentists;

(3) Is not generally available in finished form for purchase or for dispensing upon prescription;

(4) Is not offered for commercial distribution through labeling or advertising; and

(5) Is intended for use by an individual patient named in the order of a physician or dentist, and is to be made in a specific form for that patient, or is intended to meet the special needs of the physician or dentist in the course of professional practice.

(c) "FDA" means the Food and Drug Administration.

(d) "Implant" means a device that is placed into a surgically or naturally formed cavity of the human body if it is intended to remain there for a period of 30 days or more. FDA may, in order to protect public health, determine that devices placed in subjects for shorter periods are also "implants" for purposes of this part.

(e) "Institution" means a person, other than an individual, who engages in the conduct of research on subjects or in the delivery of medical services to individuals as a primary activity or as an adjunct to providing residential or custodial care to humans. The term includes, for example, a hospital, retirement home, confinement facility, academic establishment, and device manufacturer. The term has the same meaning as "facility" in section 520(g) of the act.

(f) "Institutional review board" (IRB) means any board, committee, or other group formally designated by an institution to approve and review investigations. The term has the same meaning as "institutional review committee" in section 520(g) of the act.

(g) "Investigational device" means a device, including a transitional device, that is the object of an investigation.

(h) "Investigation" means a clinical investigation or research involving one or more subjects to determine the safety or effectiveness of a device.

(i) "Investigator" means an individual who actually conducts an investigation, that is, under whose immediate direction the investigational device is administered, dispensed, or used.

(j) "Monitor," when used as a noun, means an individual designated by a sponsor or contract research organization to oversee the progress of an investigation. The monitor may be an employee of a sponsor or a consultant to the sponsor, or an employee of or consultant to a contract research organization. "Monitor," when used as a verb, means to oversee an investigation.

(k) "Noninvasive," when applied to a diagnostic device or procedure, means one that does not by design or intention: (1) Penetrate or pierce the skin or mucous membranes of the body, the ocular cavity, or the urethra, or (2) enter the ear beyond the external auditory canal, the nose beyond the nares, the mouth beyond the pharynx, the anal canal beyond the rectum, or the vagina beyond the cervical os. For purposes of this part, blood sampling that involves simple venipuncture is considered noninvasive, and the use of surplus samples of body fluids or tissues that are left over from samples taken for noninvestigational purposes is also considered noninvasive.

(l) "Person" includes any individual, partnership, corporation, association, scientific or academic establishment, Government agency or organizational unit of a Government agency, and any other legal entity.

(m) "Significant risk device" means an investigational device that:

(1) Is intended as an implant and presents a potential for serious risk to the health, safety, or welfare of a subject;

(2) Is purported or represented to be for a use in supporting or sustaining human life and presents a potential for serious risk to the health, safety, or welfare of a subject;

(3) Is for a use of substantial importance in diagnosing, curing, mitigating, or treating disease, or otherwise preventing impairment of human health and presents a potential for serious risk to the health, safety, or welfare of a subject; or

(4) Otherwise presents a potential for serious risk to the health, safety, or welfare of a subject.

(n) "Sponsor" means a person who initiates, but who does not actually conduct, the investigation, that is, the investigational device is administered, dispensed, or used under the immediate direction of another individual. A person other than an individual that uses one or more of its own employees to conduct an investigation that it has initiated is a sponsor, not a sponsor-investigator, and the employees are investigators.

(o) "Sponsor-investigator" means an individual who both initiates and actually conducts, alone or with others, an investigation, that is, under whose immediate direction the investigational device is administered, dispensed, or used. The term does not include any person other than an individual. The obligations of a sponsor-investigator under this part include those of an investigator and those of a sponsor.

(p) "Subject" means a human who participates in an investigation, either as an individual on whom or on whose specimen an investigational device is used or as a control. A subject may be in normal health or may have a medical condition or disease.

(q) "Termination" means a discontinuance, by sponsor or by withdrawl of IRB or FDA approval, of an investigation before completion.

(r) "Transitional device" means a device subject to section 520(l) of the act, that is, a device that FDA considered to be a new drug or an antibiotic drug before May 28, 1976.

(s) "Unanticipated adverse device effect" means any serious adverse effect on health or safety or any life-threatening problem or death caused by, or associated with, a device, if that effect, problem, or death was not previously identified in nature, severity; or degree of incidence in the investigational plan or application (including a supplementary plan or application), or any other unanticipated serious problem associated with a device that relates to the rights, safety, or welfare of subjects.

§ 812.5 Labeling of investigational devices.

(a) *Contents.* An investigational device or its immediate package shall bear a label with the following information: The name and place of

Federal Register / Vol. 45, No. 13 / Friday, January 18, 1980 / Rules and Regulations   3753

business of the manufacturer, packer, or distributor (in accordance with § 801.1); the quantity of contents if appropriate; all relevant contraindications, hazards, adverse effects, interfering substances or devices, warnings, and precautions; and the following statement: "CAUTION—Investigational device. Limited by Federal (or United States) law to investigational use."

(b) *Prohibitions.* The labeling of an investigational device shall not bear any statement that is false or misleading in any particular and shall not represent that the device is safe or effective for the purposes for which it is being investigated.

(c) *Animal research.* An investigational device shipped solely for research on or with laboratory animals shall bear on its label the following statement: "CAUTION—Device for investigational use in laboratory animals or other tests that do not involve human subjects."

§ 812.7 **Prohibition of promotion and other practices.**

A sponsor, investigator, or any person acting for or on behalf of a sponsor or investigator shall not:

(a) Promote or test market an investigational device, until after FDA has approved the device for commercial distribution.

(b) Commercialize an investigational device by charging the subjects or investigators for a device a price larger than that necessary to recover costs of manufacture, research, development, and handling.

(c) Unduly prolong an investigation. If data developed by the investigation indicate in the case of a class III device that premarket approval cannot be justified or in the case of a class II device that it will not comply with an applicable performance standard or an amendment to that standard, the sponsor shall promptly terminate the investigation.

(d) Represent that an investigational device is safe or effective for the purposes for which it is being investigated.

§ 812.10 **Waivers.**

(a) *Request.* A sponsor may request FDA to waive any requirement of this part. A waiver request, with supporting documentation, may be submitted separately or as part of an application to the address in § 812.19.

(b) *FDA action.* FDA may by letter grant a waiver of any requirement that FDA finds is not required by the act and is unnecessary to protect the rights, safety, or welfare of human subjects.

(c) *Effect of request.* Any requirement shall continue to apply unless and until FDA waives it.

§ 812.18 **Import and export requirements.**

(a) *Imports.* In addition to complying with other requirements of this part, a person who imports or offers for importation an investigational device subject to this part shall be the agent of the foreign exporter with respect to investigations of the device and shall act as the sponsor of the clinical investigation, or ensure that another person acts as the agent of the foreign exporter and the sponsor of the investigation.

(b) *Exports.* A person exporting an investigational device subject to this part shall obtain FDA's prior approval, as required by section 801(d) of the act.

§ 812.19 **Address for IDE correspondence.**

All applications, supplemental applications, reports, requests for waivers, requests for import or export approval, and other correspondence relating to matters covered by this part shall be addressed to the Bureau of Medical Devices, Document Control Center (HFK–20), Food and Drug Administration, 8757 Georgia Ave., Silver Spring, MD 20910. The outside wrapper of each submission shall state what the submission is, for example an "IDE application," a "supplemental IDE application," or "correspondence concerning an IDE (or an IDE application)."

Subpart B—Application and Administrative Action

§ 812.20 **Application.**

(a) *Submission.* (1) A sponsor shall submit an application to FDA if the sponsor intends to use a significant risk device in an investigation or if FDA notifies the sponsor that an application is required for an investigation.

(2) A sponsor shall submit an investigational plan and a report of prior investigations to an IRB for approval, and obtain IRB approval, before submitting an application to FDA for approval. If no IRB exists or if FDA finds that an IRB's review is inadequate, a sponsor may submit an application to FDA without first seeking and obtaining an IRB's approval of the investigation; in such a case, however, FDA may refuse to approve the application without prior IRB approval of the investigation if FDA finds that IRB review is necessary to assure protection of the rights, safety, or welfare of subjects.

(3) A sponsor shall not begin an investigation for which FDA's approval

of an application is required until FDA has approved the application. If more than one IRB must approve an investigation, and these approvals occur after submission of an application to FDA, a sponsor shall submit a supplemental application under § 812.35(b) to FDA following each additional IRB approval. A sponsor shall not begin an investigation or part of an investigation at an institution until the IRB and FDA both have approved the application or supplemental application relating to the investigation or part of an investigation at that institution.

(4) A sponsor shall submit three copies of a signed "Application for an Investigational Device Exemption" (IDE application), together with accompanying materials, by registered mail or by hand to the address in § 812.19. Subsequent correspondence concerning an application or a supplemental application shall be submitted by registered mail or by hand.

(b) *Contents.* An IDE application shall include, in the following order:

(1) The name and address of the sponsor.

(2) A complete report of prior investigations of the device and an accurate summary of those sections of the investigational plan described in § 812.25(a) through (e) or, in lieu of the summary, the complete plan. The sponsor shall submit to FDA a complete investigational plan and a complete report of prior investigations of the device if no IRB has reviewed them, if FDA has found an IRB's review inadequate, or if FDA requests them.

(3) A description of the methods, facilities, and controls used for the manufacture, processing, packing, storage, and, where appropriate, installation of the device, in sufficient detail so that a person generally familiar with good manufacturing practices can make a knowledgeable judgment about the quality control used in the manufacture of the device.

(4) An example of the agreements to be entered into by all investigators to comply with investigator obligations under this part, and a list of the names and addresses of all investigators who have signed the agreement.

(5) A certification that all investigators who will participate in the investigation have signed the agreement, that the list of investigators includes all the investigators participating in the investigation, and that no investigators will be added to the investigation until they have signed the agreement.

(6) A list of the name, address, and chairperson of each IRB that has been or will be asked to review the investigation and a certification of the action

3754 Federal Register / Vol. 45, No. 13 / Friday, January 18, 1980 / Rules and Regulations

concerning the investigation taken by each such IRB.

(7) The name and address of any institution at which a part of the investigation may be conducted that has not been identified in accordance with paragraph (b)(6) of this section.

(8) If the device is to be sold, the amount to be charged and an explanation of why sale does not constitute commercialization of the device.

(9) An environmental analysis report meeting the requirements of Part 25, when requested by FDA.

(10) Copies of all labeling for the device.

(11) Copies of all forms and informational materials to be provided to subjects to obtain informed consent.

(12) Any other relevant information FDA requests for review of the application.

(c) *Additional information.* FDA may request additional information concerning an investigation or revision in the investigational plan. The sponsor may treat such a request as a disapproval of the application for purposes of requesting a hearing under Part 16.

(d) *Information previously submitted.* Information previously submitted to the Bureau of Medical Devices in accordance with this chapter ordinarily need not be resubmitted, but may be incorporated by reference.

§ 812.25 **Investigational plan.**

The investigational plan shall include, in the following order:

(a) *Purpose.* The name and intended use of the device and the objectives and duration of the investigation.

(b) *Protocol.* A written protocol describing the methodology to be used and an analysis of the protocol demonstrating that the investigation is scientifically sound.

(c) *Risk analysis.* A description and analysis of all increased risks to which subjects will be exposed by the investigation; the manner in which these risks will be minimized; a justification for the investigation; and a description of the patient population, including the number, age, sex, and condition.

(d) *Description of device.* A description of each important component, ingredient, property, and principle of operation of the device and of each anticipated change in the device during the course of the investigation.

(e) *Monitoring procedures.* The sponsor's written procedures for monitoring the investigation and the name and address of any monitor.

(f) *Labeling.* Copies of all labeling for the device.

(g) *Consent materials.* Copies of all forms and informational materials to be provided to subjects to obtain informed consent.

(h) *IRB information.* A list of the names, locations, and chairpersons of all IRB's that have been or will be asked to review the investigation, and a certification of any action taken by any of those IRB's with respect to the investigation.

(i) *Other institutions.* The name and address of each institution at which a part of the investigation may be conducted that has not been identified in paragraph (h) of this section.

(j) *Additional records and reports.* A description of records and reports that will be maintained on the investigation in addition to those prescribed in Subpart G.

§ 812.27 **Report of prior investigations.**

(a) *General.* The report of prior investigations shall include reports of all prior clinical, animal, and laboratory testing of the device and shall be comprehensive and adequate to justify the proposed investigation.

(b) *Specific contents.* The report also shall include:

(1) A bibliography of all publications, whether adverse or supportive, that are relevant to an evaluation of the safety or effectiveness of the device, copies of all published and unpublished adverse information, and, if requested by an IRB or FDA, copies of other significant publications.

(2) A summary of all other unpublished information (whether adverse or supportive) in the possession of, or reasonably obtainable by, the sponsor that is relevant to an evaluation of the safety or effectiveness of the device.

(3) If information on nonclinical tests is provided, a statement that all nonclinical tests have been conducted in compliance with applicable requirements in the good laboratory practice regulations in Part 58, or e² detailed description of, and justification for, all differences between the practices used in the tests and those required by Part 58. Failure or inability to comply with this requirement does not justify failure to provide information on a relevant nonclinical test.

§ 812.30 **FDA action on applications.**

(a) *Approval or disapproval.* FDA will notify the sponsor in writing of the date it receives an application. FDA may approve an investigation as proposed, approve it with modifications, or disapprove it. An investigation may not begin until:

(1) Thirty days after FDA receives the application at the address in § 812.19 for the investigation of a device other than a banned device, unless FDA notifies the sponsor that the investigation may not begin; or

(2) FDA approves, by order, an IDE for the investigation.

(b) *Grounds for disapproval or withdrawal.* FDA may disapprove or withdraw approval of an application if FDA finds that:

(1) There has been a failure to comply with any requirement of this part or the act, any other applicable regulation or statute, or any condition of approval imposed by an IRB or FDA.

(2) The application or a report contains an untrue statement of a material fact, or omits material information required by this part.

(3) The sponsor fails to respond to a request for additional information within the time prescribed by FDA.

(4) There is reason to believe that the risks to the subjects are not outweighed by the anticipated benefits to the subjects and the importance of the knowledge to be gained, or informed consent is inadequate, or the investigation is scientifically unsound, or if there is reason to believe that the device is ineffective, or it is otherwise unreasonable to begin or to continue the investigation owing to the way in which the device is used or the inadequacy of:

(i) The report of prior investigations or the investigational plan;

(ii) The methods, facilities, and controls used for the manufacturing, processing, packaging, storage, and, where appropriate, installation of the device; or

(iii) Monitoring and review of the investigation.

(5) There is reason to believe that the device, as used in the investigation, is ineffective.

(c) *Notice of disapproval or withdrawal.* If FDA disapproves an application or propose to withdraw approval of an application, FDA will notify the sponsor in writing.

(1) A disapproval order will contain a complete statement of the reasons for disapproval and a statement that the sponsor has an opportunity to request a hearing under Part 16.

(2) A notice of a proposed withdrawal of approval will contain a complete statement of the reasons for withdrawal and a statement that the sponsor has an opportunity to request a hearing under Part 16. FDA will provide the opportunity for hearing before withdrawal of approval, unless FDA determines in the notice that continuation of testing under the exemption will result in an unreasonble

risk to the public health and orders withdrawal of approval before any hearing.

§ 812.35 Supplemental applications.

(a) *Changes in investigational plan.* A sponsor shall (1) submit to FDA a supplemental application if the sponsor or an investigator proposes a change in the investigational plan that may affect its scientific soundness or the rights, safety, or welfare of subjects, and (2) obtain IRB and FDA approval of the change before implementation. These requirements do not apply in the case of a deviation from the investigational plan to protect the life or physical well-being of a subject in an emergency, which deviation shall be reported to FDA within 5 working days after the sponsor learns of it.

(b) *New institutions.* A sponsor shall submit to FDA a supplemental application if an IRB other than one whose approval as a part of an investigation is certified in an application is to participate in the investigation. If the investigation is otherwise unchanged, the supplemental application shall consist of a certification of IRB approval, and updating of the information required by § 812.20(b), and a description of any modifications in the investigational plan required by the IRB as a condition of approval. A sponsor may not begin a part of an investigation at an institution until the IRB has approved the investigation and FDA has approved the supplemental application relating to that part of the investigation.

§ 812.38 Confidentiality of data and information.

(a) *Existence of IDE.* FDA will not disclose the existence of an IDE unless its existence has previously been publicly disclosed or acknowledged, until FDA approves an application for premarket approval of the device subject to the IDE; or a notice of completion of a product development protocol for the device has become effective.

(b) *Availability of summaries or data.* (1) FDA will make publicly available, upon request, a detailed summary of information concerning the safety and effectiveness of the device that was the basis for an order approving, disapproving, or withdrawing approval of an application for an IDE for a banned device. The summary shall include information on any adverse effect on health caused by the device.

(2) If a device is a banned device or if the existence of an IDE has been publicly disclosed or acknowledged, data or information contained in the file

is not available for public disclosure before approval of an application for premarket approval or the effective date of a notice of completion of a product development protocol except as provided in this section. FDA may, in its discretion, disclose a summary of selected portions of the safety and effectiveness data, that is, clinical, animal, or laboratory studies and tests of the device, for public consideration of a specific pending issue.

(3) If the existence of an IDE file has not been publicly disclosed or acknowledged, no data or information in the file are available for public disclosure except as provided in paragraphs (b)(1) and (c) of this section.

(c) *Reports of adverse effects.* Upon request or on its own initiatives, FDA shall disclose to an individual on whom an investigational device has been used a copy of a report of adverse device effects relating to that use.

(d) *Other rules.* Except as otherwise provided in this section, the availability for public disclosure of data and information in an IDE file shall be handled in accordance with § 314.14, which concerns the confidentiality of data and information in new drug applications, until the effective date of regulations concerning the confidentiality of data and information in applications for premarket approval of devices.

Subpart C—Responsibilities of Sponsors

§ 812.40 General responsibilities of sponsors.

Sponsors are responsible for selecting qualified investigators and providing them with the information they need to conduct the investigation properly, ensuring proper monitoring of the investigation, ensuring that IRB review and approval are obtained, submitting an IDE application to FDA, and ensuring that any reviewing IRB and FDA are promptly informed of significant new information about an investigation. Additional responsibilities of sponsors are described in Subparts B and G.

§ 812.43 Selecting investigators and monitors.

(a) *Selecting investigators.* A sponsor shall select investigators qualified by training and experience to investigate the device.

(b) *Control of device.* A sponsor shall ship investigational devices only to qualified investigators participating in the investigation.

(c) *Obtaining agreements.* A sponsor shall obtain from each participating

investigator a signed agreement that includes:

(1) The investigator's curriculum vitae.

(2) Where applicable, a statement of the investigator's relevant experience, including the dates, location, extent, and type of experience.

(3) If the investigator was involved in an investigation or other research that was terminated, an explanation of the circumstances that led to termination.

(4) A statement of the investigator's commitment to (i) conduct the investigation in accordance with the agreement, the investigational plan, this part and other applicable FDA regulations, and conditions of approval imposed by the reviewing IRB or FDA; (ii) supervise all testing of the device involving human subjects; and (iii) ensure that the requirements for obtaining informed consent are met.

(d) *Selecting monitors.* A sponsor shall select monitors qualified by training and experience to monitor the investigational study in accordance with this part and other applicable FDA regulations.

§ 812.45 Informing investigators.

A sponsor shall supply all investigators participating in the investigation with copies of the investigational plan and the report of prior investigations of the device.

§ 812.46 Monitoring investigations.

(a) *Securing compliance.* A sponsor who discovers that an investigator is not complying with the signed agreement, the investigational plan, the requirements of this part or other applicable FDA regulations, or any conditions of approval imposed by the reviewing IRB or FDA shall promptly either secure compliance, or discontinue shipments of the device to the investigator and terminate the investigator's participation in the investigation. A sponsor shall also require such an investigator to dispose of or return the device, unless this action would jeopardize the rights, safety, or welfare of a subject.

(b) *Unanticipated adverse device effects.* (1) A sponsor shall immediately conduct an evaluation of any unanticipated adverse device effect.

(2) A sponsor who determines that an unanticipated adverse device effect presents an unreasonable risk to subjects shall terminate all investigations or parts of investigations presenting that risk as soon as possible. Termination shall occur not later than 5 working days after the sponsor makes this determination and not later than 15 working days after the sponsor first received notice of the effect.

3756   Federal Register / Vol. 45, No. 13 / Friday, January 18, 1980 / Rules and Regulations

(c) *Resumption of terminated studies.* If the device is a significant risk device, a sponsor may not resume a terminated investigation without IRB and FDA approval. If the device is not a significant risk device, a sponsor may not resume a terminated investigation without IRB approval and, if the investigation was terminated under paragraph (b)(2), FDA approval.

Subpart D—Responsibilities of Institutional Review Boards

§ 812.60 General responsibilities of institutional review boards.

The principal responsibility of an IRB is to protect the rights, safety, and welfare of human subjects by making an initial judgment that a proposed investigation is acceptable, after reviewing the risks and benefits to human subjects, the knowledge to be gained, and the adequacy of informed consent. An IRB expresses local community attitudes and ethical standards when reviewing proposed investigations. An IRB also is responsible for ensuring that an investigation is conducted in a manner consistent with institutional policies, applicable law, and standards of professional practice. Additional responsibilities of IRB's for recordkeeping and reporting are described in Subpart G.

§ 812.62 Membership.

(a) *General.* An IRB shall be composed of not fewer than five individuals sufficiently qualified through maturity, experience, expertise, and diversity of background to ensure broad respect for its advice for safeguarding the rights, safety, and welfare of human subjects.

(b) *Diversity.* An IRB shall include at least one licensed physician, one nonphysician scientist, and one individual whose primary activities are in a nonscientific field, e.g., a lawyer, member of the clergy, ethicist, or consumer.

(c) *Nonaffiliated member.* An IRB shall include at least one member whose only affiliation with the institution is IRB membership.

(d) *Specific qualifications.* In addition to possessing the professional competence necessary to understand an investigation, the IRB as a whole shall be able to ascertain the acceptability of an investigation in terms of institutional commitments and regulations, applicable law, standards of professional conduct and practice, community attitudes, and ethical standards.

(e) *Scientific or technical knowledge.* An IRB shall have among its members or shall obtain by means of nonvoting consultants sufficient scientific and technical knowledge and expertise to be able to review proposed investigations in order to determine that the rights, safety, and welfare of human subjects are adequately protected.

(f) *Prohibition.* No IRB, institution, or other person may permit an investigator or sponsor to participate in the selection of members of an IRB that will review an investigation conducted or sponsored by that investigator or sponsor.

§ 812.65 Responsibilities and procedures.

(a) *Requirements.* An IRB shall:

(1) Adopt and follow written procedures for conducting its review of investigational plans and reports of prior investigations and for reporting its findings to the institution, investigator, and, where appropriate, the sponsor.

(2) Conduct business by a quorum of not less than a majority of the members of the IRB physically present. Regardless of the number of members physically present, a quorum shall include at least one licensed physician, one nonphysician scientist, and one member whose primary activities are in a nonscientific field.

(3) Ensure that any member having a conflict of interest, as determined by the IRB, relating to a particular investigation does not participate in the review of that investigation. This requirement does not prohibit a member from furnishing information requested by an IRB.

(4) In a timely manner review and approve, approve with modifications, disapprove, suspend, or withdraw approval of an investigation for any reason the IRB considers appropriate. An IRB may not approve an investigation unless it has determined that risks to subjects have been minimized to the extent possible consistent with the purposes of the investigation. An IRB shall disapprove or withdraw approval of an investigation if the risks to the subject are not outweighed by the anticipated benefits to the subjects and the importance of the knowledge to be gained, or if informed consent is inadequate, or if the investigation is scientifically unsound, or if there is reason to believe the device is ineffective, or if it is otherwise unreasonable, unsafe, improper, or not in the best interests of the institution to begin or to continue an investigation owing to the way in which the device is used or for any of the reasons set forth in § 812.30(b)(4)(i) through (iii).

(5) Notify the investigator and, where appropriate, the sponsor of each

decision it makes about the investigation and the basis for its decision. If an IRB determines that an investigation presented for approval under § 812.2(b)(1)(ii) involves a significant risk device, it shall so notify the investigator and, where appropriate, the sponsor. A sponsor may not begin the investigation except as provided in § 812.30(a).

(6) Continue to review an investigation that the IRB has approved until the investigation is completed or terminated. Such review shall be undertaken at intervals appropriate to the degree of risk, but not less than once a year.

(b) *Additional information.* An IRB may request an investigator or sponsor to submit additional information concerning an investigation.

(c) *Independence.* The decision of one IRB does not preclude a different decision on the same investigation by another IRB.

§ 812.70 Review of IRB actions.

Institutional officials may review and approve or reject actions by an IRB, but may not overrule an IRB disapproval; suspension; withdrawal of approval; a modification of an investigational plan determined by the IRB to be necessary or desirable to protect the rights, safety, or welfare of human subjects; or a finding that a device is a significant risk device.

Subpart E—Responsibilities of Investigators

§ 812.100 General responsibilities of investigators.

An investigator is responsible for ensuring that an investigation is conducted according to the signed agreement, the investigational plan and applicable FDA regulations, for protecting the rights, safety, and welfare of subjects under the investigator's care, and for the control of devices under investigation. An investigator also is responsible for ensuring that informed consent is obtained in accordance with Subpart F. Additional responsibilities of investigators are described in Subpart G.

§ 812.110 Specific responsibilities of investigators.

(a) *Awaiting approval.* An investigator may determine whether potential subjects would be interested in participating in an investigation, but shall not request the written informed consent of any subject to participate, and shall not allow any subject to participate before obtaining IRB and FDA approval.

(b) *Compliance.* An investigator shall conduct an investigation in accordance with the signed agreement with the sponsor, the investigational plan, this part and other applicable FDA regulations, and any conditions of approval imposed by an IRB or FDA.

(c) *Supervising device use.* An investigator shall permit an investigational device to be used only with subjects under the investigator's supervision. An investigator shall not supply an investigational device to any person not authorized under this part to receive it.

(d) *Disposing of device.* Upon completion or termination of a clinical investigation or the investigator's part of an investigation, or at the sponsor's request, an investigator shall return to the sponsor any remaining supply of the device or otherwise dispose of the device as the sponsor directs.

Subpart F—Informed Consent

§ 812.120   General.

Informed consent is a critical element of subject protection in the conduct of an investigation. The requirements in this subpart are designed to ensure that subjects understand fully the nature of potential risks and benefits of an investigation to aid them in making a voluntary choice whether to participate in it.

§ 812.122   Requirement.

(a) Before including any individual as a subject in an investigation, an investigator shall obtain from the individual, without coercion, deception, or undue influence, informed consent to participate in the investigation.

(b) Except as provided in paragraph (c), in an investigation of a significant risk device informed consent shall be evidenced by a document that includes the elements specified in § 812.130 and that is signed by the subject or, if the subject lacks legal capacity, the subject's legal representative.

(c) In the following case, an investigator may obtain informed consent by reading to the subject or the subject's legal representative a document that includes the elements specified in § 812.130 and obtaining and documenting oral informed consent:

(1) If the investigation is of a device other than a significant risk device and is subject to § 812.2(b)(1), or

(2) If a significant risk device is to be used on a subject who lacks legal capacity and whose legal representative is illiterate.

§ 812.123   Exception.

Informed consent is required, unless:

(a) The investigator determines and documents that:

(1) A life-threatening situation exists that necessitates the use of the investigational device;

(2) There is no effective alternative to use of the device;

(3) It is not feasible to obtain informed consent from the subject; and

(4) There is not sufficient time to obtain such consent from the subject's legal representative.

(b) The investigator obtains concurrence of a licensed physician not involved in the investigation, unless immediate use of the device is necessary to save the life of the subject and there is not sufficient time to obtain the concurrence of a physician.

§ 812.130   Elements of informed consent.

(a) *Requirements.* In seeking informed consent, an investigator shall provide to a subject, or to the subject's legal representative, information that includes:

(1) An explanation of the procedures to be followed, including an explanation of each procedure that is experimental.

(2) An explanation of the nature of the investigational device and an explanation of the expected duration and purpose of the use of the investigational device.

(3) A description of any attendant discomforts and risks reasonably to be expected.

(4) An explanation of likely results should the procedures fail.

(5) A description of any benefits to the subject or others reasonably to be expected.

(6) A disclosure of any appropriate alternative procedures that might be advantageous to the subject.

(7) A description of the scope of the investigation, including the number of subjects involved.

(8) An offer to answer any inquiries concerning the investigation.

(9) A disclosure that the subject, or the subject's legal representative, is free to decline participation in the investigation or to withdraw consent and to discontinue participation at any time without prejudice to the subject.

(10) A disclosure that the investigational device is being used for research purposes.

(b) *Prohibitions.* An informed consent document shall not include language that waives, or appears to waive, any of the subject's legal rights or releases the institution, its agents, the sponsor, or the investigator from liability for negligence.

Subpart G—Records and Reports

§ 812.140   Records.

(a) *Investigator records.* A participating investigator shall maintain the following accurate, complete, and current records relating to the investigator's participation in an investigation:

(1) All correspondence with another investigator, an IRB, the sponsor, a monitor, or FDA, including required reports.

(2) Records of receipt, use or disposition of a device that relate to:

(i) The type and quantity of the device, the dates of its receipt, and the batch number or code mark.

(ii) The names of all persons who received, used, or disposed of each device.

(iii) Why and how many units of the device have been returned to the sponsor, repaired, or otherwise disposed of.

(3) Records of each subject's case history and exposure to the device. Such records shall include:

(i) Documents evidencing informed consent and, for any use of a device by the investigator without informed consent, any written concurrence of a licensed physician and a brief description of the circumstances justifying the failure to obtain informed consent.

(ii) All relevant observations, including records concerning adverse device effects (whether anticipated or unanticipated), information and data on the condition of each subject upon entering, and during the course of, the investigation, including information about relevant previous medical history and the results of all diagnostic tests.

(iii) A record of the exposure of each subject to the investigational device, including the date and time of each use, and any other therapy.

(4) The protocol, with documents showing the dates of and reasons for each deviation from the protocol.

(5) Any other records that FDA requires to be maintained by regulation or by specific requirement for a category of investigations or a particular investigation.

(b) *Sponsor records.* A sponsor shall maintain the following accurate, complete, and current records relating to an investigation:

(1) All correspondence with another sponsor, a monitor, an investigator, an IRB, or FDA, including required reports.

(2) Records of shipment and disposition. Records of shipment shall include the name and address of the consignee, type and quantity of device, date of shipment, and batch number or

code mark. Records of disposition shall describe the batch number or code marks of any devices returned to the sponsor, repaired, or disposed of in other ways by the investigator or another person, and the reasons for and method of disposal.

(3) Signed investigator agreements.

(4) For each investigation subject to § 812.2(b)(1) of a device other than a significant risk device, the records described in paragraph (b)(5) of this section and the following records, consolidated in one location and available for FDA inspection and copying:

(i) The name and intended use of the device and the objectives of the investigation;

(ii) A brief explanation of why the device is not a significant risk device;

(iii) The name and address of each investigator;

(iv) The name and address of each IRB that has reviewed the investigation;

(v) A statement of the extent to which the good manufacturing practice regulation in Part 820 will be followed in manufacturing the device and a copy of any quality assurance program that is followed with respect to the device; and

(vi) Any other information required by FDA.

(5) Records concerning adverse device effects (whether anticipated or unanticipated) and complaints and

(6) Any other records that FDA requires to be maintained by regulation or by specific requirement for a category of investigation or a particular investigation.

(c) *IRB records.* A reviewing IRB shall maintain the following accurate, complete, and current records relating to that IRB's review of an investigation:

(1) All correspondence with another IRB, an investigator, a sponsor, a monitor, or FDA.

(2) Records of the membership of the IRB and of its members' employment relationship with the institution with which the IRB is associated, for example full-time or part-time employee, member of governing panel or board, paid or unpaid consultant. Such records shall include:

(i) Each member's name, earned degrees (if any), position or occupation, specialty field (if any), representative capacity, and other pertinent indications of qualifications, such as board certifications or licenses;

(ii) Each member's employment or other relationship with an investigator or sponsor whose investigation is reviewed by the IRB, for example full-time or part-time employee, member of governing panel or board, paid or unpaid consultant; and

(3) Minutes of attendance at each meeting and of each decision concerning an investigation.

(d) *Retention period.* Records required by this subpart shall be maintained during the investigation and for a period of 2 years after the latter of the following two dates: The date on which the investigation is terminated or completed, or the date that the records are no longer required for purposes of supporting a premarket approval application or a notice of completion of a product development protocol.

(e) *Records custody.* An investigator, sponsor, or IRB may withdraw from the responsibility to maintain records for the period required in paragraph (d) of this section and transfer custody of the records to any other person who will accept responsibility for them under this part, including the requirements of § 812.145. Notice of a transfer shall be given to FDA not later than 10 working days after transfer occurs.

§ 812.145 Inspections.

(a) *Entry and inspection.* A sponsor or an investigator who has authority to grant access shall permit authorized FDA employees, at reasonable times and in a reasonable manner, to enter and inspect any establishment where devices are held (including any establishment where devices are manufactured, processed, packed, installed, used, or implanted or where records of results from use of devices are kept).

(b) *Records inspection.* A sponsor, IRB, or investigator, or any other person acting on behalf of such a person with respect to an investigation, shall permit authorized FDA employees, at reasonable times and in a reasonable manner, to inspect and copy all records relating to an investigation.

(c) *Records identifying subjects.* An investigator shall permit authorized FDA employees to inspect and copy records that identify subjects, upon notice that FDA has reason to suspect that adequate informed consent was not obtained, or that reports required to be submitted by the investigator to the sponsor or IRB have not been submitted or are incomplete, inaccurate, false, or misleading.

§ 812.150 Reports.

(a) *Investigator reports.* An investigator shall prepare and submit the following complete, accurate, and timely reports:

(1) *Unanticipated adverse device effects.* An investigator shall submit to the sponsor and to the reviewing IRB a report of any unanticipated adverse device effect occurring during an

investigation as soon as possible, but in no event later than 10 working days after the investigator first learns of the effect.

(2) *Withdrawal of IRB approval.* An investigator shall report to the sponsor, within 5 working days, a withdrawal of approval by the reviewing IRB of the investigator's part of an investigation.

(3) *Progress.* An investigator shall submit progress reports on the investigation to the sponsor, the monitor, and the reviewing IRB at regular intervals, but in no event less often than yearly.

(4) *Deviations from the investigational plan.* An investigator shall notify the sponsor and the reviewing IRB of any deviation from the investigational plan to protect the life or physical well-being of a subject in an emergency. Such notice shall be given as soon as possible, but in no event later than 5 working days after the emergency occurred. Except in such an emergency, prior approval by the sponsor is required for changes in or deviations from a plan, and, if these changes or deviations may affect scientific soundness of the plan or the rights, safety, or welfare of human subjects, IRB and FDA approval under § 812.35(a) also is required.

(5) *Informed consent.* If an investigator uses a device without obtaining informed consent, the investigator shall report such use to the sponsor and the reviewing IRB within 5 working days after the use occurs.

(6) *Final report.* An investigator shall, within 3 months after termination or completion of the investigation or the investigator's part of the investigation, submit a final report to the sponsor and the reviewing IRB.

(7) *Other.* An investigator shall, upon request by a reviewing IRB or FDA, provide accurate, complete, and current information about any aspect of the investigation.

(b) *Sponsor reports.* A sponsor shall prepare and submit the following complete, accurate, and timely reports:

(1) *Unanticipated adverse device effects.* A sponsor who conducts an evaluation of an unanticipated adverse device effect under § 812.46(b) shall report the results of such evaluation to FDA and to all reviewing IRB's and participating investigators within 10 working days after the sponsor first receives notice of the effect. Thereafter the sponsor shall submit such additional reports concerning the effect as FDA requests.

(2) *Withdrawal of IRB approval.* A sponsor shall notify FDA and all reviewing IRB's and participating investigators of any withdrawal of

Federal Register / Vol. 45, No. 13 / Friday, January 18, 1980 / Rules and Regulations   3759

approval of an investigation or a part of an investigation by a reviewing IRB within 5 working days after receipt of the withdrawal of approval.

(3) *Withdrawal of FDA approval.* A sponsor shall notify all reviewing IRB's and participating investigators of any withdrawal of FDA approval of the investigation, and shall do so within 5 working days after receipt of notice of the withdrawal of approval.

(4) *Current investigator list.* A sponsor shall submit to FDA, at 6-month intervals, a current list of the names and addresses of all investigators participating in the investigation. The sponsor shall submit the first such list 6 months after FDA approval.

(5) *Progress reports.* At regular intervals, but in no event less often than yearly, a sponsor shall submit progress reports to FDA and to all reviewing IRB's.

(6) *Recall and device disposition.* A sponsor shall notify FDA and all reveiwing IRB's of any request that an investigator return, repair, or otherwise dispose of any units of a device. Such notice shall occur within 30 working days after the request is made and shall state why the request was made.

(7) *Final report.* In the case of a significant risk device, the sponsor shall notify FDA within 30 working days of the completion or termination of the investigation and shall submit a final report to FDA and all reviewing the IRB's and participating investigators within 6 months after completion or termination. In the case of a device that is not a significant risk device, the sponsor shall submit a final report to all reviewing IRB's within 6 months after termination or completion.

(8) *Informed consent.* A sponsor shall submit to FDA a copy of any report by an investigator under paragraph (a)(5) of this section of use of a device without obtaining informed consent, within 5 working days of receipt of notice of such use.

(9) *Significant risk device determinations.* If an IRB determines that a device is a significant risk device, and the sponsor had proposed that the IRB consider the device not to be a significant risk device, the sponsor shall submit to FDA a report of the IRB's determination within 5 working days after the sponsor first learns of the IRB's determination.

(10) *Other.* A sponsor shall, upon request by a reviewing IRB or FDA, provide accurate, complete, and current information about any aspect of the investigation.

Effective date. The reporting and recordkeeping requirements contained in this rule have been submitted for approval by the Office of Management and Budget (OMB) in accordance with the Federal Reports Act of 1942. This regulation will become effective July 16, 1980, provided that approval of the OMB is received by that date. If OMB does not approve, without change, the reporting and recordkeeping requirements contained in the rule, FDA will revise the rule as necessary to comply with the decision of OMB. FDA will publish a notice in a future issue of the **Federal Register** concerning OMB's decision on these requirements.

(Secs. 301, 501, 502, 520, 701(a), 702, 704, 801, 52 Stat. 1042–1043 as amended, 1049–1051 as amended, 1055, 1056–1058 as amended, 67 Stat. 476–477 as amended, 90 Stat. 565–574, (21 U.S.C. 331, 351, 352, 360, 371(a), 372, 374, 381))

Dated: January 8, 1980.

Jere E. Goyan,

Commissioner of Food and Drugs.

[FR Doc. 80-1256 Filed 1–17–80; 8.45 a.m.]

BILLING CODE 4110-03-M

FEDERAL FORMS

<table>
<tr><td>DEPARTMENT OF HEALTH, EDUCATION, AND WELFARE
PUBLIC HEALTH SERVICE
FOOD AND DRUG ADMINISTRATION
5600 FISHERS LANE
ROCKVILLE, MARYLAND 20852</td><td>STATEMENT OF INVESTIGATOR</td><td>Form Approved
OMB No. 57-R0029</td></tr>
</table>

TO: SUPPLIER OF DRUG (Name and address, include Zip Code)	NAME OF INVESTIGATOR (Print or Type)
	DATE
	NAME OF DRUG

Dear Sir:

The undersigned, _____ submits this statement as required by section 505(1) of the Federal Food, Drug, and Cosmetic Act and § 312.1 of Title 21 of the Code of Federal Regulations as a condition for receiving and conducting clinical investigations with a new drug limited by Federal (or United States) law to investigational use.

1.   STATEMENT OF EDUCATION AND EXPERIENCE

a. COLLEGES, UNIVERSITIES, AND MEDICAL OR OTHER PROFESSIONAL SCHOOLS ATTENDED, WITH DATES OF ATTENDANCE, DEGREES, AND DATES DEGREES WERE AWARDED

b. POSTGRADUATE MEDICAL OR OTHER PROFESSIONAL TRAINING (Indicate dates, names of institutions, and nature of training)

c. TEACHING OR RESEARCH EXPERIENCE (Indicate dates, institutions, and brief description of experience)

d. EXPERIENCE IN MEDICAL PRACTICE OR OTHER PROFESSIONAL EXPERIENCE (Indicate dates, institutional affiliations, nature of practice, or other professional experience)

e. REPRESENTIVE LIST OF PERTINENT MEDICAL OR OTHER SCIENTIFIC PUBLICATIONS (Indicate titles of articles, names of publications and volume, page number, and date)

FD FORM 1573 (7/75)   PREVIOUS EDITION MAY BE USED UNTIL SUPPLY IS EXHAUSTED.

259

2a. If the investigation is to be conducted on institutionalized subjects or is conducted by an individual affiliated with an institution which agrees to assume responsibility for the study, assurance must be given that an institutional review committee is responsible for initial and continuing review and approval of the proposed clinical study. The membership must be comprised of sufficient members of varying background, that is, lawyers, clergymen, or laymen as well as scientists, to assure complete and adequate review of the research project. The membership must possess not only broad competence to comprehend the nature of the project, but also other competencies necessary to judge the acceptability of the project or activity in terms of institutional regulations, relevant law, standards of professional practice, and community acceptance. Assurance must be presented that the investigator has not participated in the selection of committee members; that the review committee does not allow participation in its review and conclusions by any individual involved in the conduct of the research activity under review (except to provide information to the committee); that the investigator will report to the committee for review any emergent problems, serious adverse reactions, or proposed procedural changes which may affect the status of the investigation and that no such change will be made without committee approval except where necessary to eliminate apparent immediate hazards; that reviews of the study will be conducted by the review committee at intervals appropriate to the degree of risk, but not exceeding 1 year, to assure that the research project is being conducted in compliance with the committee's understanding and recommendations; that the review committee is provided all the information on the research project necessary for its complete review of the project; and that the review committee maintains adequate documentation of its activities and develops adequate procedures for reporting its findings to the institution. The documents maintained by the committee are to include the names and qualifications of committee members, records of information provided to subjects in obtaining informed consent, committee discussion on substantive issues and their resolution, committee recommendations, and dated reports of successive reviews as they are performed. Copies of all documents are to be retained for a period of 3 years past the completion or discontinuance of the study and are to be made available upon request to duly authorized representatives of the Food and Drug Administration. (Favorable recommendations by the committee are subject to further appropriate review and rejection by institution officials. Unfavorable recommendations, restrictions, or conditions may not be overruled by the institution officials.) Procedures for the organization and operation of institutional review committees are contained in guidelines issued pursuant to Chapter 1-40 of the Grants Administration Manual of the U.S. Department of Health, Education, and Welfare, available from the U.S. Government Printing Office. It is recommended that these guidelines be followed in establishing institutional review committees and that the committees function according to the procedures described therein. A signing of the Form FD 1573 will be regarded as providing the above necessary assurances, however, if the institution has on file with the Department of Health, Education, and Welfare, Division of Research Grants, National Institutes of Health, an "accepted general assurance," and the same committee is to review the proposed study using the same procedures, this is acceptable in lieu of the above assurances and a statement to this effect should be provided with the signed FD 1573. (In addition to sponsor's continuing responsiblity to monitor the study, the Food and Drug Administration will undertake investigations in institutions periodically to determine whether the committees are operating in accord with the assurances given by the sponsor.)

b. A description of any clinical laboratory facilities that will be used. (If this information has been submitted to the sponsor and reported by him on Form FD 1571, reference to the previous submission will be adequate).

---

3. *The investigational drug will be used by the undersigned or under his supervision in accordance with the plan of investigation described as follows: (Outline the plan of investigation including approximation of the number of subjects to be treated with the drug and the number to be employed as controls, if any; clinical uses to be investigated; characteristics of subjects by age, sex and condition; the kind of clinical observations and laboratory tests to be undertaken prior to, during, and after administration of the drug; the estimated duration of the investigation; and a description or copies of report forms to be used to maintain an adequate record of the observations and test results obtained. This plan may include reasonable alternates and variations and should be supplemented or amended when any significant change in direction or scope of the investigation is undertaken.)*

4. THE UNDERSIGNED UNDERSTANDS THAT THE FOLLOWING CONDITIONS, GENERALLY APPLICABLE TO NEW DRUGS FOR INVESTIGATIONAL USE, GOVERN HIS RECEIPTS AND USE OF THIS INVESTIGATIONAL DRUG:

a. The sponsor is required to supply the investigator with full information concerning the preclinical investigations that justify clinical trials, together with fully informative material describing any prior investigations and experience and any possible hazards, contraindications, side-effects, and precautions to be taken into account in the course of the investigation.

b. The investigator is required to maintain adequate records of the disposition of all receipts of the drug, including dates, quantities, and use by subjects, and if the investigation is terminated, suspended, discontinued, or completed, to return to the sponsor any unused supply of the drug. If the investigational drug is subject to the Comprehensive Drug Abuse Prevention and Control Act of 1970, adequate precautions must be taken including storage of the investigational drug in a securely locked, substantially constructed cabinet, or other securely locked substantially constructed enclosure, access to which is limited, to prevent theft or diversion of the substance into illegal channels of distribution.

c. The investigator is required to prepare and maintain adequate and accurate case histories designed to record all observations and other data pertinent to the investigation on each individual treated with the drug or employed as a control in the investigation.

d. The investigator is required to furnish his reports to the sponsor of the drug who is responsible for collecting and evaluating the results obtained by various investigators. The sponsor is required to present progress reports to the Food and Drug Administration at appropriate intervals not exceeding 1 year. Any adverse effect that may reasonably be regarded as caused by, or probably caused by, the new drug shall be reported to the sponsor promptly, and if the adverse effect is alarming, it shall be reported immediately. An adequate report of the investigation should be furnished to the sponsor shortly after completion of the investigation.

e. The investigator shall maintain the records of disposition of the drug and the case histories described above for a period of 2 years following the date a new-drug application is approved for the drug; or if the application is not approved, until 2 years after the investigation is discontinued. Upon the request of a scientifically trained and properly authorized employee of the Department, at reasonable times, the investigator will make such records available for inspection and copying. The subjects' names need not be divulged unless the records of particular individuals require a more detailed study of the cases, or unless there is reason to believe that the records do not represent actual cases studied, or do not represent actual results obtained.

f. The investigator certifies that the drug will be administered only to subjects under his personal supervision or under the supervision of the following investigators responsible to him,

_____

_____

and that the drug will not be supplied to any other investigator or to any clinic for administration to subjects.

g. The investigator certifies that he will inform any subjects including subjects used as controls, or their representatives, that drugs are being used for investigational purposes, and will obtain the consent of the subjects, or their representatives, except where this is not feasible or, in the investigator's professional judgment, is contrary to the best interests of the subjects.

h. The investigator is required to assure the sponsor that for investigations involving institutionalized subjects, the studies will not be initiated until the institutional review committee has reviewed and approved the study. (The organization and procedure requirements for such a committee should be explained to the investigator by the sponsor as set forth in Form FD 1571, division 10, unit c.

Very truly yours,

_____

(Name of Investigator)

_____

(Address)

_____

(This form should be supplemented or amended from time to time if new subjects are added or if significant changes are made in the plan of investigation.)

DEPARTMENT OF HEALTH, EDUCATION, AND WELFARE	☐ GRANT ☐ CONTRACT ☐ FELLOW ☐ OTHER
PROTECTION OF HUMAN SUBJECTS ASSURANCE/CERTIFICATION/DECLARATION ☐ ORIGINAL ☐ FOLLOWUP ☐ REVISION	☐ NEW ☐ RENEWAL ☐ CONTINUATION APPLICATION IDENTIFICATION NUMBER *(if known)*

STATEMENT OF POLICY: Safeguarding the rights and welfare of subjects at risk in activities supported under grants and contracts from DHEW is primarily the responsibility of the institution which receives or is accountable to DHEW for the funds awarded for the support of the activity. In order to provide for the adequate discharge of this institutional responsibility, it is the policy of DHEW that no activity involving human subjects to be supported by DHEW grants or contracts shall be undertaken unless the Institutional Review Board has reviewed and approved such activity, and the institution has submitted to DHEW a certification of such review and approval, in accordance with the requirements of Public Law 93-348, as implemented by Part 46 of Title 45 of the Code of Federal Regulations, as amended, (45 CFR 46). Administration of the DHEW policy and regulation is the responsibility of the Office for Protection from Research Risks, National Institutes of Health, Bethesda, Md 20014.

1. TITLE OF PROPOSAL OR ACTIVITY

2. PRINCIPAL INVESTIGATOR/ACTIVITY DIRECTOR/FELLOW

3. DECLARATION THAT HUMAN SUBJECTS EITHER WOULD OR WOULD NOT BE INVOLVED

☐ A. NO INDIVIDUALS WHO MIGHT BE CONSIDERED HUMAN SUBJECTS, INCLUDING THOSE FROM WHOM ORGANS, TISSUES, FLUIDS, OR OTHER MATERIALS WOULD BE DERIVED, OR WHO COULD BE IDENTIFIED BY PERSONAL DATA, WOULD BE INVOLVED IN THE PROPOSED ACTIVITY. (IF NO HUMAN SUBJECTS WOULD BE INVOLVED, CHECK THIS BOX AND PROCEED TO ITEM 7. PROPOSALS DETERMINED BY THE AGENCY TO INVOLVE HUMAN SUBJECTS WILL BE RETURNED.)

☐ B. HUMAN SUBJECTS WOULD BE INVOLVED IN THE PROPOSED ACTIVITY AS EITHER: ☐ NONE OF THE FOLLOWING, OR INCLUDING: ☐ MINORS, ☐ FETUSES, ☐ ABORTUSES, ☐ PREGNANT WOMEN, ☐ PRISONERS, ☐ MENTALLY RETARDED, ☐ MENTALLY DISABLED. UNDER SECTION 6. COOPERATING INSTITUTIONS, ON REVERSE OF THIS FORM, GIVE NAME OF INSTITUTION AND NAME AND ADDRESS OF OFFICIAL(S) AUTHORIZING ACCESS TO ANY SUBJECTS IN FACILITIES NOT UNDER DIRECT CONTROL OF THE APPLICANT OR OFFERING INSTITUTION.

4. DECLARATION OF ASSURANCE STATUS/CERTIFICATION OF REVIEW

☐ A. THIS INSTITUTION HAS NOT PREVIOUSLY FILED AN ASSURANCE AND ASSURANCE IMPLEMENTING PROCEDURES FOR THE PROTECTION OF HUMAN SUBJECTS WITH THE DHEW THAT APPLIES TO THIS APPLICATION OR ACTIVITY. ASSURANCE IS HEREBY GIVEN THAT THIS INSTITUTION WILL COMPLY WITH REQUIREMENTS OF *DHEW Regulation 45 CFR 46*, THAT IT HAS ESTABLISHED AN INSTITUTIONAL REVIEW BOARD FOR THE PROTECTION OF HUMAN SUBJECTS AND, WHEN REQUESTED, WILL SUBMIT TO DHEW DOCUMENTATION AND CERTIFICATION OF SUCH REVIEWS AND PROCEDURES AS MAY BE REQUIRED FOR IMPLEMENTATION OF THIS ASSURANCE FOR THE PROPOSED PROJECT OR ACTIVITY.

☐ B. THIS INSTITUTION HAS AN APPROVED GENERAL ASSURANCE (DHEW ASSURANCE NUMBER _____)OR AN ACTIVE SPECIAL ASSURANCE FOR THIS ONGOING ACTIVITY, ON FILE WITH DHEW. THE SIGNER CERTIFIES THAT ALL ACTIVITIES IN THIS APPLICATION PROPOSING TO INVOLVE HUMAN SUBJECTS HAVE BEEN REVIEWED AND APPROVED BY THIS INSTITUTION'S INSTITUTIONAL REVIEW BOARD IN A CONVENED MEETING ON THE DATE OF _____ IN ACCORDANCE WITH THE REQUIREMENTS OF THE *Code of Federal Regulations on Protection of Human Subjects (45 CFR 46)*. THIS CERTIFICATION INCLUDES, WHEN APPLICABLE, REQUIREMENTS FOR CERTIFYING FDA STATUS FOR EACH INVESTIGATIONAL NEW DRUG TO BE USED (SEE REVERSE SIDE OF THIS FORM).

THE INSTITUTIONAL REVIEW BOARD HAS DETERMINED, AND THE INSTITUTIONAL OFFICIAL SIGNING BELOW CONCURS THAT:

EITHER ☐ HUMAN SUBJECTS WILL NOT BE AT RISK;   OR ☐ HUMAN SUBJECTS WILL BE AT RISK.

5. AND 6. SEE REVERSE SIDE

7. NAME AND ADDRESS OF INSTITUTION

8. TITLE OF INSTITUTIONAL OFFICIAL	TELEPHONE NUMBER
SIGNATURE OF INSTITUTIONAL OFFICIAL	DATE

HEW-596 (Rev 4-75)

ENCLOSE THIS FORM WITH THE PROPOSAL OR RETURN IT TO REQUESTING AGENCY.

5. INVESTIGATIONAL NEW DRUGS - ADDITIONAL CERTIFICATION REQUIREMENT

SECTION 46.17 OF TITLE 45 OF THE Code of Federal Regulations states, "Where an organization is required to prepare or to submit a certification . . . and the proposal involves an investigational new drug within the meaning of The Food, Drug, and Cosmetic Act, the drug shall be identified in the certification together with a statement that the 30-day delay required by 21 CFR 130.3(a)(2) has elapsed and the Food and Drug Administration has not, prior to expiration of such 30-day interval, requested that the sponsor continue to withhold or to restrict use of the drug in human subjects, or that the Food and Drug Administration has waived the 30-day delay requirement; provided, however, that in those cases in which the 30-day delay interval has neither expired nor been waived, a statement shall be forwarded to DHEW upon such expiration or upon receipt of a waiver. No certification shall be considered acceptable until such statement has been received."

INVESTIGATIONAL NEW DRUG CERTIFICATION

TO CERTIFY COMPLIANCE WITH FDA REQUIREMENTS FOR PROPOSED USE OF INVESTIGATIONAL NEW DRUGS IN ADDITION TO CERTIFICATION OF INSTITUTIONAL REVIEW BOARD APPROVAL, THE FOLLOWING REPORT FORMAT SHOULD BE USED FOR EACH IND: (ATTACH ADDITIONAL IND CERTIFICATIONS AS NECESSARY).

- IND FORMS FILED ☐ FDA 1571, ☐ FDA 1572, ☐ FDA 1573

- NAME OF IND AND SPONSOR _____

- DATE OF 30-DAY EXPIRATION OR FDA WAIVER
  (FUTURE DATE REQUIRES FOLLOWUP REPORT TO AGENCY) _____

- FDA RESTRICTION _____

- SIGNATURE OF INVESTIGATOR _____ DATE _____

6. COOPERATING INSTITUTIONS - ADDITIONAL REPORTING REQUIREMENT

SECTION 46.16 OF TITLE 45 OF THE Code of Federal Regulations IMPOSES SPECIAL REQUIREMENTS ON THE CONDUCT OF STUDIES OR ACTIVITIES IN WHICH THE GRANTEE OR PRIME CONTRACTOR OBTAINS ACCESS TO ALL OR SOME OF THE SUBJECTS THROUGH COOPERATING INSTITUTIONS NOT UNDER ITS CONTROL. IN ORDER THAT THE DHEW BE FULLY INFORMED, THE FOLLOWING REPORT IS REQUESTED WHEN APPLICABLE.

USE FOLLOWING REPORT FORMAT FOR EACH INSTITUTION OTHER THAN GRANTEE OR CONTRACTING INSTITUTION WITH RESPONSIBILITY FOR HUMAN SUBJECTS PARTICIPATING IN THIS ACTIVITY (ATTACH ADDITIONAL REPORT SHEETS AS NECESSARY).

INSTITUTIONAL AUTHORIZATION FOR ACCESS TO SUBJECTS

- SUBJECTS: STATUS (WARDS, RESIDENTS, EMPLOYEES, PATIENTS, ETC.) _____

  NUMBER _____ AGE RANGE _____
  NAME OF OFFICIAL (PLEASE PRINT) _____
  TITLE _____ TELEPHONE _____

  NAME AND ADDRESS OF _____
  COOPERATING INSTITUTION _____
  _____

- OFFICIAL SIGNATURE _____

NOTES: (e.g., report of modification in proposal as submitted to agency affecting human subjects involvement)

APPLICATION FOR APPROVAL OF A RESEARCH PROJECT

LIJ-HMC Institutional Review Board (IRB) for the Protection of Human Subjects Participating in Research

Instructions to PIs: Complete Section I and II. Complete either A, B, or C as appropriate to your project. Investigators are referred to the Research Manual which contains institutional policies for the protection of human subjects participating in research. The Office of Grants Management will be pleased to provide any assistance.

I. Investigator _____Dept. or Address _____
    Co-Investigator_____Telephone Extension: _____

    Title of Protocol: _____
    _____

    Dept. Head Approval _____
    Other Dept. Involved: Yes_____ No _____   Other Institution Involved: Yes _____ No _____

    If Yes, Dept. Head Approval_____
    Name of Other Institution: _____

☐ A. RESEARCH PRESENTING POSSIBLE RISK TO SUBJECTS: e.g. drug and medical device trials, surgical and other invasive procedures, studies involving randomization, placebo controls, etc. Please submit:

   1. Nine (9) copies of the complete protocol;

   2. Nine (9) copies of a lay summary of the project — i.e. an explanation of the study in *non-medical terminology.*

   3. Nine (9) copies of a properly executed consent form. Note: Investigator must circle the appropriate cost statement on page 2.

☐ B. RESEARCH PRESENTING MINIMAL RISK TO SUBJECTS: In order for your study to be categorized as a "minimal risk" project it must fall into one or more of the following areas. Please indicate the category:

   ☐ 1. Collection of hair and nail clipping, excreta and external secretions, uncannulated saliva, placenta removed at delivery, amniotic fluid at the time of rupture of the membrane, deciduous teeth, and permanent teeth if patient care indicates a need for extraction. Collection of dental plaque and calculus done in a noninvasive manner performed according to standard prophylactic techniques.

   ☐ 2. Collection of blood samples by venipuncture, in amounts not exceeding 450 milliliters in an 8 week period, and no more often than twice a week, from subjects over 18 years of age, in good health, and not pregnant.

   ☐ 3. Recording of data from subjects 18 years or older using noninvasive procedures routinely employed in clinical practice (e.g. weighing, testing sensory acuity, electrocardiography, electroencephalography, thermography — NOT X-RAYS OR MICROWAVES).

   ☐ 4. Moderate exercise by healthy volunteers.

   ☐ 5. Voice recordings made for research purposes.

   ☐ 6. Research on behavior: ☐ perception studies; ☐ cognition; ☐ game theory; ☐ test development, where the investigator does not manipulate subjects' behavior and the research will not involve stress to subjects.

Attach one (1) copy of the complete protocol and a summary in non-medical terminology which will be given to the subject.

FOR NO RISK RESEARCH — See Next Page

COMPLETE SECTION II ON NEXT PAGE

☐ C. RESEARCH PRESENTING NO RISK TO SUBJECTS: In order for your research to be considered as a "No Risk" study, it must fall into one or more of the following categories. Please indicate which area(s) apply:

   ☐ 1. Use of educational tests for which there is no subject identifying data.

   ☐ 2. Research involving collection or study of charts, specimens, or medical records for which there will be no subject identifying data.

   ☐ 3. Research involving questionnaires, surveys, interviews. Subjects cannot be identified from data; subjects responses, if known, will not place them at risk; research does not deal with sensitive aspects of subjects behavior (e.g. illegal conduct, drug or alcohol use, sexual behavior). *All of the conditions must be met.*

Attach one (1) copy of the protocol and, if applicable, a summary of the project in non-medical terminology which will be provided to the subject.

II. ALL INVESTIGATORS MUST SIGN THE FOLLOWING STATEMENT OF ASSURANCE.

The proposed investigation involves the use of human subjects. I am submitting this form with a description of my project, prepared in accordance with institutional policy for the protection of human subjects participating in research. I understand the Medical Center's policy concerning research involving human subjects and I agree:

   1. to obtain informed consent of subjects who are to participate in this project;

   2. to report to the Human Subjects Review Committee any unanticipated effects on subjects which become apparent during the course or as a result of experimentation and the actions taken as a result;

   3. to cooperate with members of the Committee charged with the continuing review of this project;

   4. to obtain prior approval from the Committee before amending or altering the scope of the project or implementing changes in the approved consent form;

   5. to maintain documentation of consent forms and progress reports as required by institutional policy.

Signature _____ Date _____

Note: Investigators are referred to the Research Manual for complete statement of institutional policy and procedures regarding research with human subjects.

III. FOR COMPLETION BY THE OFFICE OF GRANTS MANAGEMENT

Disposition of the Protocol: _____ Date of Committee Meeting _____

                                                 Subcommittee _____

☐ In accordance with institutional policy, this protocol was approved via the procedures for expedited review.

_____   _____
HSRC Staff Committee   Date   Chairman/   Date
                                    Subcommittee Chairman

Continuing Review Scheduled for_____ .

Sample Consent Form

NAME OF INSTITUTION

Investigator(s) or Project Director:

Title of Protocol:

Expected Duration of Study:

I agree to (have my ward) participate as a subject in the following project:

I understand that the project will include the following experimental procedures:

I understand that the possible discomforts or risks may be as follows:

I also understand that the possible and desired benefits of my participation in this study are:

I am aware that the following alternative procedures could be of benefit to me (my ward):

The Institution will make available hospital facilities and professional attention at
                                    or at an affiliated hospital should the research study have been
conducted at an affiliated hospital to a patient who may suffer physical injury resulting directly from the
research. The expense for hospitalization and professional attention will be borne by the patient. Financial
compensation from   will not be provided.

I have been given the opportunity to ask further questions and know that I can do so during the course of
the project. I understand that in comparison to the costs of medical care which I would normally bear,
my participation in this study may: a) greatly increase; b) increase; c) have no effect; d) decrease;
e) greatly decrease my cost. (Note: *The researcher must circle one of the preceding choices.)*

I am aware that I am under no obligation to participate in this project. I am also aware that I may
withdraw my (ward's) participation at any time without prejudice to my medical treatment at

I understand that although the FDA may inspect records which reveal my participation in this study,
my identity and participation will be kept confidential to the extent permitted by law.

I further understand that should I have any questions about my (ward's) treatment or any other matter
relative to my (ward's) participation in this project, I may call (title of responsible administrator and
telephone number) and I will be given the opportunity to discuss, in confidence, any questions with a
member of the Institutional Review Board for the Protection of Human Subjects. This is a Committee
which, as required by Federal regulations and State Law, is an independent Committee composed of
physicians, administrative staff and lay members of the community not affiliated with this institution.
This Committee has evaluated the potential risks and possible benefits of this study.

A copy of this consent form has been offered to me.

Patient's Name_____ /Signature _____Date_____

Parent/Guardian's Name_____ /Signature _____Date_____

Witness: _____ /Signature _____

Relationship of Witness to Patient _____

IRB Reviewer's Checklist

Instructions for Reviewers: The attached research protocol places human subjects at risk. Please review the protocol, lay summary and consent form. Complete this form and return it to the Coordinator of the Human Subjects Review Committee.

Principal Investigator   Dept./School

Title of Protocol

Reviewer   Sub-Committee

Choose one:

☐   This protocol should be approved in the form presented.

☐   This protocol should be approved with changes and/or deletions as indicated below under number(s)

☐   Should be disapproved.

SECTION A: Risk/Benefit Analysis

1.   What specific benefit(s) will a subject derive from participating in this project?

2.   What specific risks will the subject undergo?

3.   Does the benefit to the subject justify the risks of the proposed treatment?

4.   What would be the effects to the subject's health or well-being of not rendering the proposed treatment?

5.   If risk to subject is minimal, what indirect benefits will he obtain from participation in this subject?

SECTION B: Analysis of Protocol

1.   Does the protocol contain provision for development of complete medical profile (detailed medical history, tests for allergic and/or adverse reactions, etc.)?
     ☐ Yes   ☐ No   Comments:

2.   Does the protocol provide for the development of a comprehensive psychological/social profile?
     ☐ Yes   ☐ No   Comments:

3. Are the subjects for this project minors, pregnant women or mental incompetents?
☐ Yes   ☐ No   Comments:

(a) If yes, does the protocol contain an indepth examination of the possible psychological effects and risks of participation in this project?
☐ Yes   ☐ No   Comments:

(b) If yes, does the protocol provide for an adequate examination procedure?
☐ Yes   ☐ No   Comments:

(c) If yes, do the potential benefits justify the subjects participation?
☐ Yes   ☐ No   Comments:

4. Does the protocol contain adequate provision for the protection of subject's privacy and anonymity in its handling of data (i.e. processing, storage, retrieval, etc.)?
☐ Yes   ☐ No   Comments:

5. Does the protocol, in its recruitment procedures, indicate any responsibility of undue influence on subjects to participate?
☐ Yes   ☐ No   Comments:

SECTION C: Analysis of Consent Form

1. Does the proposed subject consent form, in an adequate manner and *in layman's language,* contain:

(a) A fair explanation of the procedures to be followed, including an identification of those which are experimental?
☐ Yes   ☐ No   Comments:

(b) A description of the attendant discomforts and risks?
☐ Yes   ☐ No   Comments:

(c) A description of the benefits to be expected?
☐ Yes   ☐ No   Comments:

(d) A disclosure of appropriate alternative procedures that would be advantageous for the subject?
☐ Yes   ☐ No   Comments:

2. Does the proposed subject consent form contain any language which, in any way, would seem to waive a subject's legal rights against the Medical Center, its agents, or the practitioners conducting the experiment, form liability for negligence?
☐ Yes   ☐ No   Comments:

Additional Comments:

_____
Signature   Date

REQUEST FOR INVESTIGATIONAL DRUG

Name Plate of Patient

```
┌─────────────────────────┐
│                         │
│                         │
└─────────────────────────┘
```

INSTRUCTIONS: Complete Part 1 in research situations, non-emergency clinical situations, and emergency/life threatening situations and send to Pharmacy. In situations where the patient or family cannot sign the consent form, fill out Parts I and II. In *all* emergency/life threatening situations, send copies to Departmental Chairman, Pharmacy (attach consent form and protocol) and the Grants Management Office.

___

Part I

1. ☐   Inpatient   Indicate Hospital Division _____

   ☐   Outpatient

2. Generic Name:

3. Tentative Brand Name and/or Company:

4. Dosage Schedule:

5. Route of Administration:

6. Informed consent of patient has been obtained.   ☐ Yes   ☐   No. Complete Part II. Attach explanation as may be necessary.

7. Protocol Number _____ or brief description of use:

8. Authorized Investigator _____ Date _____
                                          (Signature)

9. Departmental Chairman Approval _____ Date _____
                                          (Signature)

___

Part II   TO BE COMPLETED IN SITUATIONS WHERE PATIENT OR FAMILY CANNOT PROVIDE INFORMED CONSENT.

              Send completed form with copy of protocol to Pharmacy.

10. Administrative Approval: _____ Date _____
                                          (Signature)

___

PHARMACY USE ONLY:

Amount Dispensed _____ Lot Number_____

Dispensed by:_____ R. Ph. _____ Date _____

INVESTIGATIONAL DRUG FACT SHEET

To familiarize nursing and pharmacy personnel with this investigational drug, the principal investigator is requested to complete this form, forwarding it to the Director of Pharmacy Services. A copy of this form will be inserted in each nursing unit's Research Manual.

N.B. In accordance with institutional policy, Registered Nurses may only administer an investigational drug if they have received detailed *written* directions from the physician authorized to prescribe the drug.

1. Name of Drug (Designation to be used in prescribing and labeling) or Title of Protocol:

2. Synonyms (Generic/Trade Names):

3. Preparations (Dosage forms and strengths available from the pharmacy):

4. Usual Therapeutic Dose:

5. Possible Dosage Range:

6. Route of Administration:

7. Indications:

8. Expected Therapeutic Effect:

9. Possible Untoward Effect:

10. Contraindications:

11. Antidotes:

12. Drug Storage:

13. Special Instructions for Preparation and Administration:

14. Investigator(s) (Indicate Principal Investigator by P.I., List Phone Extension):

15. DATE:
_____

Please use space below for any additional pertinent information:

LONG ISLAND JEWISH-HILLSIDE MEDICAL CENTER

RESEARCH INCIDENT REPORT

Date_____

Title of Protocol

Investigator(s)   Department_____

Subject Data

Name_____ Chart #_____

Address_____ City_____ State_____

Telephone_____

Age_____ Date of Birth_____ Sex_____ Marital Status_____ Inpt_____ Outpt_____

Study Status: Experimental____ Control ____ Date of Subject's Entry Into Study_____

Date of Incident_____ Location_____

Personnel Involved_____ _____

Incident

Supportive Medical Information (Lab. Tests, X-rays, etc.)

SAMPLE CONSENT FORMS FOR

SPECIFIC AREAS OF RESEARCH

Sample Consent Form
Research Involving Children

Participant's Name: Date

Project Title: Haemophilus Vaccine   Protocol No.

Expected Duration of Project:

DESCRIPTION AND EXPLANATION OF PROCEDURE: *Haemophilus influenzae* is the most common cause of bacterial meningitis in infants and young children. It is also a common cause of other serious infections including blood stream infections, pneumonia, and joint infections. More rarely, it may cause middle ear infection.

Physicians have previously tried to make a vaccine against this bacterium but were not able to get young infants to respond. We would now like to test and compare three new vaccine preparations which appear to be more effective in producing an immune response in very young children.

If you agree to participate, your child will receive one of these three vaccines in addition to the regular well-baby shots (diphtheria, pertussin and tetanus, or DPT) which are given at 2, 4, and 6 months of age.

However, 1 in 4 babies will receive an injection of an ineffective salt solution, called a placebo, instead of the new vaccine. This is to compare the effectiveness of the new vaccines. Neither you nor your doctor will know whether your child is receiving one of the 3 vaccines or the salt solution. This is called a random trial.

The *Haemophilus* vaccine (or placebo) will be injected into the right thigh muscle and the DPT vaccine into the left thigh muscle. Your child will be monitored for any side effects of the immunization. A small amount of blood (one teaspoon or less) will be obtained from a vein or by heelstick before each shot and at seven, nine, and twelve months of age in order to measure the protective activity produced by the vaccines. At twelve months of age, all children will receive a shot of the vaccine preparation found to be most effective in the first part of the study.

Finally, a throat and rectal swab may be obtained during some of the visits to look for the presence of *Haemophilus influenzae* bacteria.

RISKS AND DISCOMFORTS: Similar vaccines made against *Haemophilus influenzae* have been tested in over 50,000 children and infants and no serious side effects have been noted. The new vaccines used in this study have also been tested in 50 to 200 children without serious side effects. A few children have had fever and some have had slight redness or swelling at the injection site. These side effects have been less severe than those with the routine DPT shots. The risks of drawing blood are minimal. They include some discomfort to the child, an occasional bruise at the site of skin puncture and, extremely rarely, an infection at the site. The cultures cause minimal discomfort.

POTENTIAL BENEFITS: The potential benefits of this study are that we may be able to protect your child from infections caused by *Haemophilus influenzae*. In addition, the results of this and similar studies may eventually lead to the introduction of this vaccine for routine use in all children. All vaccines, cultures, lab tests and pediatric visits during this study will be provided at no cost to you.

275

ALTERNATIVES: If you do not participate in this study, your child will receive only the routine well-baby shots.

I have fully explained to (parent/guardian) the nature and purpose of the above-described procedure and the risks involved in its performance. I have answered and will answer all questions to the best of my ability. I will inform the participant of any changes in the procedure or the risks and benefits if any should occur during or after the course of the study.

Doctor's signature   Date

CONSENT: I have been satisfactorily informed of the above-described procedure with its possible risks and benefits. I give permission for my child's participation in this study. I know that Dr.   or his associates will be available to answer any questions I may have. If I feel my questions have not been adequately answered, I may request to speak to a member of the Institutional Review Board by calling (name of responsible administrator and telephone number). I understand that I am free to withdraw this consent and discontinue participation in this project at any time, even after signing this form, and it will not affect my/my child's care. I have been offered a copy of this form.

Signature of parent/guardian   date

Witness to signatures   date

Sample Consent Form
Cancer Research

Title of Research Protocol: EST 1479 - Acute Non-lymphocytic Leukemia. A
                              Randomized Trial of Intensive Consolidation Therapy

I, (name of participant), willingly agree to participate in the above treatment
study, explained to me by Dr.   , as part of the Oncology Group studies
approved by the National Cancer Institute and (name of institution).

NATURE OF DISEASE AND PURPOSE OF STUDY: It has been explained to me that I have
acute non-lymphocytic leukemia. I understand that the investigational study
described below has been undertaken in an effort to: 1) to slow, stop or decrease
the growth of my leukemia; 2) to gain information about my disease; and 3)
evaluate the effectiveness in my disease of three drugs which have been shown to
be effective in some patients with this disease.

DESCRIPTION OF STUDY: The chemotherapy that I will receive if I choose to participate in this study consists of Daunomycin and Cytosine Arabinoside given by vein,
plus 6-Thioguanine given by mouth. I will receive all three of these drugs during
the first three to six weeks of treatment. After that, if my body shows a good
response to treatment, I will either receive two more courses of the same three
drugs before starting weekly treatment with a two-drug combination of Cytosine
Arabinoside plus 6-Thioguanine or I will start the two-drug weekly treatment alone.
It is not clear at the present time whether the two extra courses of the three-
drug treatment will help to maintain my response.

POTENTIAL RISKS: Chemotherapy (drug treatment) has a number of side effects.
These side effects may include nausea and vomiting, diarrhea, lowering of the blood
count, increased risk of serious infections, bleeding, and hair loss. In some
instances these side effects may be so severe that they may cause an individual to
die. Sometimes, people who are treated with these drugs experience skin rash,
fever, jaundice or heart failure. There is the possibility of genetic risk to a
baby if patients on these treatments become pregnant. Also there is always the
risk of very uncommon or previously unknown side effects occurring. My doctor will
watch for all side effects and will adjust my treatment when necessary. I understand that there may be added costs for any drugs which are used to treat my side
effects.

POTENTIAL BENEFITS: It is not possible to predict whether any personal benefit
will result from the use of the drug treatments, but it is understood that if no
benefit is occurring, I will be so informed and the drug treatment program will
be stopped.

ALTERNATIVE TREATMENT(S): Alternative treatments which could be advantageous in
my case include treatments with different drugs or different combinations of drugs.
No alternatives are known to be superior to the treatment to be used here and all
have similar side effects.

The physicians involved in my care have made themselves available to answer any
questions I have concerning this program. In addition, I understand that I am
free to ask my physicians any questions concerning this program that I wish in the
future. I have been told that I should feel free to discuss my disease and the
prospect of recovery with my doctor. I have been assured that any procedures
related solely to research which would not otherwise be necessary will be explained
to me. Some of these procedures may result in added costs and some of these costs
may not be covered by insurance. My doctor will discuss these with me. In the
event that physical harm occurs as a result of the research over and above the
expected from the disease or its usual treatment, no compensation will be provided.

I understand that I am free to withdraw my consent to participate in this treatment program without prejudice to my subsequent care and to seek care from any physician of my choice at any time.

I understand that a record of my progress while on the study will be kept in a confidential form at (name of institution) and also in a computer file at the statistical headquarters of the Oncology Group. The confidentiality of the central computer record is carefully guarded and no information by which I can be identified will be released or published. Histopathologic material (slides) will be sent to a central office for review.

I have read all of the above, asked questions, received answers concerning areas I did not understand, and willingly give my consent to participate in this program.

(Patient Signature)   (Date)

(Witness Signature)   (Date)

(Physician Signature)   (Date)

Sample Consent Form
Research Involving Psychiatric Patients

Title of Protocol: A Comparison of Lithium and Imipramine in the Prevention of
                     Recurrent Depression

Principal Investigator: John M. Kane, M.D.

I hereby agree to participate as a subject in the following project: a comparison
of imipramine (Tofranil^R) and lithium carbonate (Eskalith^R) in the prevention of
depression.

I understand the project will include the following experimental procedures: the
use of lithium and imipramine for the prevention of depression is widespread.
Lithium carbonate has been approved by the FDA for the treatment of mood swings.
It has not as yet been approved for the prevention of depression alone.   Imipramine
is a standard approved drug for the treatment of depression.

I understand that I will receive either one of these medications during the course
of this project.   I understand that I will not know which medication I am receiving.
Each month examinations will be carried out by psychiatrists in order to determine
the benefit of each treatment or any possible side effects.   I understand that this
project will last for up to 3 years.   If I experience any significant side effects
from the medication I am taking, I will be taken out of the study.

I understand that I will have a series of laboratory tests to determine any possible
side effects from these medications.

I understand that the possible discomforts or risks are as follows: Imipramine
can cause dry mouth, blurred vision, dizziness, postural hypotension and electrocardiogram changes.

Lithium can produce hand tremor, muscle fatigue, weight gain, increased thirst and
urination.   In some cases, lithium can induce hypothyroidism (diminished activity
of the thyroid gland).   I understand that this can readily be treated by thyroid
replacement therapy.   Recently a question of possible kidney changes occurring
during lithium treatment has arisen.   I understand that I will be carefully monitored to determine whether or not I am experiencing any early signs of such side
effects.

I understand what the possible and desired benefits of this program are: I understand that both of these agents have been shown to be of some value in the prevention of recurrent episodes of depression.

I understand that either treatment may be effective in this regard and that the
overall purpose of this study is to determine which treatment is more effective.

I understand that I will be closely watched for any change in my clinical condition.   This study is expected to continue for at least 12 months.

The Institution will make available hospital facilities and professional attention
at (name of institution) (or at an affiliated hospital should the research study
have been conducted at the affiliated hospital) to a patient who may suffer physical
injury resulting directly from the research.   The expense for hospitalization and
professional attention will be borne by the patient.   Financial compensation from
(name of institution) will not be provided.

I understand that these medications, particularly lithium, if taken during the first trimester of pregnancy may increase the risk of fetal abnormalities. Therefore, I agree either to use adequate forms of birth control during the course of this investigation, or to discuss with my physician my desire to become pregnant before doing so.

I have been given an opportunity to ask further questions and understand that I can do so during the course of the project. I understand that in comparison to the cost of medical care which I would normally incur, my participation in this study may decrease my costs.

I am aware that I am under no obligation to participate in this project. I am also aware that I may withdraw my participation at any time without prejudice to my continued medical treatment at (name of institution).

I further understand that should I have any questions about my participation in this project, I may call (name of responsible institutional administrator and telephone number), and I will be given an opportunity to discuss in confidence, any questions with a member of the Institutional Review Board. This is a Committee which, as required by Federal Regulations and New York State law, is an independent committee composed of medical center physicians and staff as well as lay members of the community not affiliated with this institution. This Committee has evaluated the potential risks and possible benefits of the study.

I understand that although the FDA may inspect records which reveal my participation in this study, my identity and participation will be kept confidential to the extent permitted by law.

Patient's Name   Date
                    (Please print)

Patient's Signature   Chart No.

Parent/Guardian's Name   /Signature
                    (Please print)

Witness   /Signature

Relationship of Witness to Patient

Sample Consent Form
Experimental Invasive Procedure

Investigator(s) or Project Director:

Title of Protocol: Coronary Percutaneous Transluminal Angioplasty

I hereby agree to (have my ward) participate as a subject in the following project:
Use of a special balloon tip catheter (tube) to be used to dilate (widen) my
coronary arteries, using methods similar to cardiac catheterization. Investigators
wish to learn how safe and effective this procedure will be for improving blood flow
to the heart and how long this improvement will last.

DESCRIPTION OF THE RESEARCH PROJECT: Patients who are considered candidates for
coronary bypass surgery will be offered a new alternative method, presently under
investigation. Investigators will use a specially designed catheter with a balloon
to dilate the blood vessels to the heart. In order to ascertain health status,
patients will be followed up at approximately 1 year intervals for a five year period.

POSSIBLE DISCOMFORTS OR RISKS: Problems arising from this technique are similar to
those of a cardiac catheterization, with local bleeding at puncture site, which, at
times, may require surgery. Occasional patient may require emergency coronary bypass surgery. As with cardiac catheterization, there is a risk of heart attack or
death.

POTENTIAL BENEFITS OF MY PARTICIPATION: To dilate coronary arteries, which are
blocked, may be associated with desired effect of relief of chest pain. This procedure may thus negate the need for open heart surgery and both its risks and costs.
Balloon procedure will require a hospital stay of 3 to 4 days. No doctor's fees
are entailed.

I am aware that the following alternative procedures could be of benefit to me (my
ward): Open heart surgery with a bypass procedure.

The institution will make available hospital facilities and professional attention
at (name of institution) (or at an affiliated hospital should the research study
have been conducted at the affiliated hospital) to a patient who may suffer physical
injury resulting directly from the research. The expense for hospitalization and
professional attention will be borne by the patient. Financial compensation from
(name of organization) will not be provided.

I have been given the opportunity to ask further questions and know that I can do
so during the course of the project. I understand that in comparison to the costs
of medical care which I would normally bear, my participation in this study may
decrease my cost.

I am aware that I am under no obligation to participate in this project. I am
also aware that I may withdraw my (ward's) participation at any time without
prejudice to my medical treatment at (name of institution).

I understand that although the FDA may inspect records which reveal my participation in this study, my identity and participation will be kept confidential to the
extent permitted by law.

I further understand that should I have any questions about my (ward's) treatment or any other matter relative to my (ward's) participation in this project, I may call (name of responsible institutional administrator and telephone number) and I will be given an opportunity to discuss, in confidence, any questions with a member of the Institutional Review Board. This is a Committee which, as required by Federal regulations and New York State law, is an independent Committee composed of physicians and staff as well as lay members of the community not affiliated with this institution. This Committee has evaluated the potential risks and possible benefits of this study.

Patient's Name   Date
              (Please Print)

Patient's Signature   Chart No.

Parent/Guardian's Name   /Signature
                   (Please print)

Witness   /Signature

Relationship of Witness to Patient

Sample Consent Form
Research Utilizing an Investigational
Medical Device

Participant's Name: Date:

Project Title: Telimiterized Monitoring of Dwyer   Protocol No.:
                 Instrumentation

DESCRIPTION AND EXPLANATION OF PROCEDURE: You/your child has a deformity of the
spine which is getting worse and requires an operation. The best method available
at the moment is to approach the spine from the front and use a system of cables and
screws known as the Dwyer Instrumentation. This is a standard operation for certain
spine deformities. However, one of the problems with the operation is that the bones
may fail to fuse together. The reason for this failure is not always clear. In an
attempt to discover more about the forces acting on the spine after operation so
that we can decide whether you/your child needs to be kept lying down or allowed to
sit up while the spine is healing, we have made a small electronic instrument which
is attached to the cable of the Dwyer Instrumentation and put in at the time of
your/your child's operation. This small instrument, which is approximately the size
of the tip of the little finger contains no batteries or active electrical current;
but it does allow us to read the forces on the spine by a simple loop, which is
placed around you/your child at various times after the operation. By taking the
readings, we will be able to tell what is the best position for you/your child to
be kept in, how much of the body we have to put in a cast, and how long we will have
to keep it on.

This is the first time this device will be used in humans, but other similar devices
are currently in use for monitoring other kinds of spine operations. Animal
experiments with this particular device have shown that it has worked well even
when in place for as long as a year.

RISKS AND DISCOMFORTS: (1) It is not likely that you/your child will feel this
instrument has been placed in the spine, because it adds very little to what has to
be put in there anyway.
                        (2) Putting in this instrument will not make the operation
any more dangerous, but it will add about 15 minutes to the normal time of 5 to 7
hours.
                        (3) When the test readings are done after the operation,
a special loop is placed around the body. This causes no discomfort. Each set of
readings will take about ten minutes. They will be done each day you/your child
is in the hospital, and whenever you/your child returns for regular visits.
                        (4) There was no evidence of rusting or leaking in the
animal experiments. However, in the extremely unlikely event they should occur,
it might be necessary to remove the instrument. This can be done with a relatively
simple operation which will not affect the outcome of the main operation on the
spine. However, such an operation would require the use of a general anesthesia
and the normal risks of a surgical procedure.

POTENTIAL BENEFITS: The benefit to you/your child from having this procedure done
is that we will have more definite knowledge of what has happened to the spine, so
that we can tell the best treatment to use after the operation. It will also help
to tell us when we can safely remove the cast, and whether or not the operation is
likely to be successful.

As well as the direct benefits to you/your child, this system of testing forces on
the spine can be used to benefit other patients with different conditions who also
require this type of spinal surgery.

ALTERNATIVES: WE have been doing this operation without putting the electronic instrument in, and you/your child's operation could be done without. We would then have to use the ordinary method of estimating how to treat you/your child after the operation by taking repeated x-rays.

I have fully explained to (participant/parent/guardian) the nature and purpose of the above-described procedure and the risks involved in its performance. I have answered and will answer all questions to the best of my ability. I will inform the participant of any changes in the procedure or the risks and benefits if any should occur during or after the course of the study.

                              Doctor's signature   Date

CONSENT: I have been satisfactorily informed of the above-described procedure with its possible risks and benefits. I give permission for my/my child's participation in this study. I know that Dr.   or his associates will be available to answer any questions I may have. If I feel my questions have not been adequately answered, I may request to speak to a member of the Institutional Review Board by calling (name of responsible institutional administrator and telephone number). I understand that I am free to withdraw this consent and discontinue participation in this project at any time and it will not affect my/my child's care. I have been offered a copy of this form.

Witness to signatures   Date   Signature of participant/parent/guardian

Sample Consent Form
Research Involving an Investigational
New Drug

Investigator or Project Director:

Title of Protocol: Propranolol Heart Trial in the Secondary Prevention of
Coronary Heart Disease

I hereby agree to (have my ward) participate as a subject in the following project:
Use of a drug called propranolol for long term treatment of patients who have heart
attacks in order to determine whether the drug prevents future attacks or complications, and prolongs life.

DESCRIPTION OF RESEARCH: Propranolol is an accepted drug employed in treatment of
angina pectoris (chest pain brought on by exertion). Its use in myocardial infarc-
tion (heart attack) is currently being evaluated in this study. Patients will be
given capsules of the drug or a harmless substitute (placebo). Routine blood tests,
electrocardiogram and chest x-rays will be obtained for 2-4 years as indicated for
normal follow-up treatment.

POSSIBLE DISCOMFORTS OR RISKS: The possible rare side effects of propranolol
include: slow heart rate, low blood pressure, fatigue, and rash. The drug will
be discontinued if any adverse reaction occurs. A small hematoma (black and blue
mark) may occur after blood drawing.

POTENTIAL BENEFITS: The drug could protect against potentially fatal irregular
heart rhythm that might follow heart attack. It may also reduce the area of damage
to the heart during a subsequent heart attack. There will be no charge for
physicians fees or for any test obtained during the course of this study.

I am aware that the following alternative procedures could be of benefit to me
(my ward): The alternative is routine care for heart attack without the possible
preventative use of propranolol.

I have been given an opportunity to ask further questions and know that I can do
so during the course of the project.

I am aware that I may withdraw my (my ward's) participation at any time.

I understand that although the FDA may inspect records which reveal my participation in this study, my identity and participation will be kept confidential to
the extent permitted by law.

I understand that should I have any questions about my (ward's) treatment or any
other matter relative to my (ward's) participation in this project, I may call
(name of responsible institutional administrator and telephone number), and I
will be given an opportunity to discuss, in confidence, any question with a member
of the Institutional Review Board.

I also understand that under no circumstances will my name or participation in
this project be disclosed.

Patient Name   Date
              (Print)

Patient Number   If Other Than Patient
                                                               (Print Name)

Patient Signature   If Other Than Patient
                                                               (Signature)

Witness   Relationship to Patient
              (Print)

Witness Signature   Address

Address of Witness

Sample Consent Form
Research Involving an Emotional Assessment
Interview with Patients

I understand that my interview will include questions concerning my past and present
emotional state, and I agree to this procedure. The entire interview will last 90
to 120 minutes. The questions I will be asked will deal with a range of mental
health issues, such as my feelings about myself and others. I understand that the
questions I will be asked are consistent with good medical practice.

In the event I have a history of psychiatric or medical hospitalizations, I may be
asked by the clinical staff to sign a waiver to get past medical records. The information thus obtained would significantly contribute to the study. I understand,
however, that I can refuse this procedure, and that such information will not be
made available to the clinical staff without my signed consent.

I understand that this information will be coded, studied, and used for the purpose
of increasing the general understanding of mental health problems. My name will be
removed from this information and a code number will be used instead. I have been
assured that access to the information obtained in my interview will be limited to
scientific and medical staff directly engaged in this study and in my clinical care.

I understand that participating or not participating in the procedures described
above will not affect my clinical care and that I can discontinue the interview at
any point.

I give my interviewers permission to contact my relatives (parents, siblings and
children). I understand that my relatives will be asked to consent to undergo a
similar interview, and I agree to this procedure. I also understand that their
consent or refusal to participate in the study will not interfere with my care.

The investigators have offered to answer any questions I might have about the procedures described above, or about the results of the study. I understand that
there are no risks or discomfort in participating in this project. I understand
that my participation in this project may help the investigators gather data which
may be of some use in studying certain disorders in some psychiatric patients.

I further understand that should I have any further questions about my participation in this study, I may call (name of responsible institutional administrator
and telephone number) and will be given an opportunity to discuss, in confidence,
any questions with a member of the Institutional Review Board.

Patient's Name   Date
                  please print

Patient's Signature   Chart No.

Parent/Guardian's Name   /Signature
                        please print

Witness   /Signature

Relationship of Witness to Patient

Sample Consent Form
Research Involving an Emotional Assessment
Interview with Relatives of Patients

I understand that my interview will include questions concerning my past and present emotional state, and I agree to participate in this interview. The entire interview will last 90 to 120 minutes. The questions I will be asked will deal with a range of mental health issues, such as my feelings about myself and others. I understand that the questions I will be asked are consistent with good medical practice.

In the event I have a history of medical or psychiatric hospitalizations, I may be asked by the clinical staff to sign a waiver to obtain past medical records. The information thus obtained would significantly contribute to the usefulness of my interview. I understand, however, that I can refuse to permit the clinical staff to obtain my past medical records.

I understand that this information obtained from the interview will be coded, studied and used for the purpose of increasing the general understanding of mental health problems. My name will be removed from this information and a code number will be used instead. I have been assured that access to the information obtained in my interview will be limited to scientific staff directly engaged in this study and in the clinical care of my relative.

I understand that participating or not participating in the procedures described above will not affect the regular clinical care delivered to my relative.

The investigators have offered to answer any questions I might have about the procedures described above, or about the results of the study. I understand that there are no risks or discomforts in participating in this project. I understand that my participation in this project may help the investigators gather data which may be of some use in studying certain disorders in some psychiatric patients.

I further understand that should I have any further questions about my participation in this study, I may call (name of responsible institutional administrator and telephone number) and will be given an opportunity to discuss, in confidence, any question with a member of the Institutional Review Board.

Subject's Name   Date

Subject's Signature   Date

Witness   Signature
          print or type name

Index